JÁNOS BOLYAI, NON-EUCLIDEAN GEOMETRY,

AND THE NATURE OF SPACE

János Bolyai, Non-Euclidean Geometry,
and the Nature of Space

JEREMY J. GRAY

BURNDY LIBRARY PUBLICATIONS

NEW SERIES, NUMBER I

CAMBRIDGE, MASSACHUSETTS

BURNDY LIBRARY

2004

The facsimiles in this book have been reproduced from the
Burndy Library's copies of the following two books:

János Bolyai, "Appendix: scientiam spatii absolute veram
exhibens," which was published as an appendix to volume one
of Farkas Bolyai, *Tentamen juventutem studiosam in elementa
matheseos purae, elementaris ac sublimioris, methodo intuitiva,
evidentiaque huic propria, introducendi: cum appendice triplici auctore
professore matheseos et physices chemiaeque publ. ordinario.*
2 volumes. Maros Vásárhelyini: Typis Collegii Reformatorum
per Josephum et Simeonem Kali de felső Vist., 1832–1833.

János Bolyai, *The Science Absolute of Space: Independent of the Truth
or Falsity of Euclid's Axiom XI (which can never be decided a priori).*
Translated by Dr. George Bruce Halsted. Fourth edition.
Austin: The University of Texas, 1896.

All images are reproduced
from the collection of the Burndy Library,
except the photograph by Eugène Atget (page 30),
which is reproduced courtesy of the Cleveland Museum of Art.
Eugène Atget, French, 1857–1927.
Fountain at l'École Polytechnique, 1902.
Albumen print, gold toned, 17.6 x 21.5 cm. © The Cleveland
Museum of Art, 2002. John L. Severance Fund, 1985.116.

Distributed by The MIT Press, Cambridge, Massachusetts
and London, England.

CONTENTS

FOREWORD

IN 1942, when the Burndy Library was but a half decade old, the Library sponsored the publication of I. Bernard Cohen's *Ole Roemer and the Determination of the Speed of Light*. That book, which contained a facsimile of Roemer's paper as well as Professor Cohen's commentary, was the first of dozens of books issued by the Library over the next fifty years. The Burndy publications were varied in content, format, and author – many were written by the Library's founder, Bern Dibner, himself – but all shared a focus on some aspect of the Library's collection, and all were united by a desire to make that collection, and, by extension, the history of science and technology come to life for a readership that included specialists and nonspecialists alike. The idea was that the Library, a tool for scholars, should not be of use or interest only to scholars – that it should not preach solely to the choir.

In 1992 the Burndy Library moved from Norwalk, Connecticut, to the campus of the Massachusetts

Institute of Technology, in Cambridge, and began a second life. The decade since that move has been one of unprecedented growth in the Library's collection and its audience. In that spirit of renewal, the Library now launches a new series of publications. This first book in the new series parallels in many ways the first in the old: it embraces a facsimile (two facsimiles, actually) of a canonical text in the history of science, and a commentary on that text by an academic, Jeremy Gray, of the Open University, who is both a prominent scholar and a skilled explainer. We hope that the new series will be as diverse and as long-lived as the first.

Benjamin Weiss

CURATOR OF RARE BOOKS
BURNDY LIBRARY

Preclarissimū opus elementoꝝ Euclidis
mentis Campani ꝑspicacissimi in artē geo

P

Unctus est cuius
longitudo sine lati
tremitates sunt du
ē ab vno pūcto ad
sio in extremitate
piens. ¶ Superficie
dinē tm̄ habet: cui
¶ Superficies pl
am extensio in ext
¶ Angulus planu
rius cōtactus: qua
ꝑficiē applicatioq; nō directa. ¶ Quādo a
linee recte rectilineꝰ angulus noiatur. ¶
steterit duoq; anguli vtrobiq; fuerint eq̄le
¶ Lineaq; linee superstans ei cui sup̄stat pp
gulus vero qui recto maior est obtusus dic
recto acutꝰ appellat. ¶ Terminꝰ ē qd vnīu
ra ē q̄ termino vel terminis ꝯtinet. ¶ Circu
dē linea cōtenta: q̄ circūferētia noiat: ī cuiꝰ
linee recte ad circūferentiā exeūtes sibiīnu
quidē punctꝰ cētꝝ circuli dicit. ¶ Diamet
sup eiꝰ centꝝ transiēs extremitatesq; suas
circulū in duo media diuidit. ¶ Semicircu
metro circuli ⁊ medietate circūferentie cōt
est figura plana recta linea ⁊ parte circūfer
lo quidem aut maior aut minor. ¶ Rectilin
lineis cōtinent quaꝝ quedā trilatere q̄ trib
quadrilatere q̄ quatuor rectis lineis: q̄dā m

Bolyai's *Appendix:*

An Introduction

WE REMEMBER the Hungarian mathematician János Bolyai, who lived from 1802 to 1860, because he is one of the few people who changed forever our ideas about space. Space is a difficult concept. All visible things, from the objects outside your window to the stars and galaxies, have positions and shapes. They may move, they may change their shape, and when they do we see them as moving into other regions of – well, what, exactly? The natural answer is space, empty space. Of course, nearby regions of that space are filled with air, but when we allow ourselves to think of the whole universe, we seem driven to imagine all the possible positions objects can be in, including the presently unoccupied ones. They all exist in some immense, open arena, which we shall call physical space. The nature of physical space is hard to grasp, because it is so fundamental. The whole force of Bolyai's discovery was that almost every thing people had said about it could be wrong. As a result, our ideas not just about physics but also mathematics began to change, slowly at first but eventually in radical ways.

What had people believed about space? Aristotle's universe, as described in his *On the Heavens*, was a finite one. At its core was the Earth, which was enveloped in a region of change, growth, and decay. In daily experience things move, and they fall when you drop them. But beyond a certain point the universe seems very different. Beyond the moon lie the planets. The planets wander about in the sky (whence their Greek name) but don't otherwise change much, whereas the moon waxes and wanes continually. And beyond the planets are the stars, which seem always to be in the same relative positions to each other, and to wheel round in the sky as though they are embedded in a giant sphere. So Aristotle divided the universe into three regions, the sublunary sphere of daily life, the region of the planets, and a sphere of fixed stars.

His was not the only Greek view, but it became a Western orthodoxy. It was accompanied by various theories about how the planets moved, of which Ptolemy's was the best developed and which came down with numerous Arab and Islamic additions to the West. These theories aimed at describing the motion of the planets, which were supposed to rotate on various sets of spheres. The aim was to describe their motion as the composition of rotations about various centers which might themselves also move. Circular motion, at least in the heavens, was supposed to be eternal. The spheres, though, were primarily math-

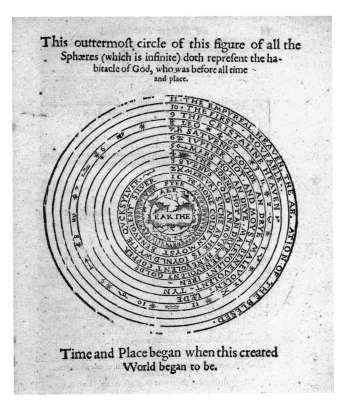

Aristotle's universe as depicted in Thomas Tymme's
1612 treatise, *A Dialogue Philosophical, Wherein
Nature's Secret Closet is Opened.*

ematical objects. There was no widely accepted, consistent model to explain the motion, though Ptolemy did have such a model in mind.

Aristotle's universe was proclaimed not long before the other Greek theory that will occupy us much more: Euclidean geometry. We do not know much

Pᷓeclariſſimū opus elementoᷓ Euclidis megareſis vna cū cō/
mentis Campani pſpicaciſſimi in arte geometriā incipit feliciᷓ.

Punctus est cuius ps non est. ℂ Linea est
longitudo ſine latitudie cuius quidem ex/
tremitates ſunt duo puncta. ℂ Linea recta
é ab vno pūcto ad aliū breuiſſima exten
ſio in extremitates ſuas vtrūqʒ eoᷓ reci/
piens. ℂ Supficies é q̄ lōgitudinē ᷓ latitu
dinē tm̄ habet: cui°termini quidē ſūt linee
ℂ Superficies plana é ab vna linea ad ali
am extenſio in extremitates ſuas recipiēs
ℂ Angulus planus é duarum lineaᷓ alte/
rius cōtactus: quaᷓ expanſio eſt ſuper ſu
pficiē applicatioqʒ nō directa. ℂ Quādo auté angulū cōtinent due
linee recte rectilineº angulus noiatur. ℂ Cū̄ recta linea ſup rectā
ſteterit duoqʒ anguli vtrobiqʒ fuerint eǭles eoᷓ vterqʒ rectºerit.
ℂ Lineaqʒ linee ſuperſtans ei cui ſupſtat ppēdicularis vocaᷓ. ℂ An
gulus vero qui recto maior eſt obtuſus dicif. ℂ Angulºvero mio
recto acutº appellaf. ℂ Terminºé qd̄ vniuſcuiuſqʒ finis é. ℂ Figu/
ra é q̄ termino vel terminis ᷓtinef. ℂ Circulºé figura plana vna q̄/
dē linea cōtenta: q̄ circūferētia noiaᷓ: i cui°medio pūctºé a quo oēs
linee recte ad circūferentiā exeūtes ſubiūnicè ſunt equales. Et bic
quidē punctºcētᷓ circuli dicif. ℂ Diameter circuli é linea recta q̄
ſup eiºcentᷓ tranſiés extremitateſqʒ ſuas circūferentie applicans
circulū in duo media diuidit. ℂ Semicirculus é figura plana dia/
metro circuli ᷓ medietate circūferentie cōtenta. ℂ Poᷓtio circuli
eſt figura plana recta linea ᷓ parte circūferentie cōtenta: ſemicircu
lo quidem aut maioᷓ aut minor. ℂ Rectilinee figure ſūt que rectis
lineis cōtinenf quaᷓ quedā trilatere q̄ tribºrectis lineis: quedam
quadrilatere q̄ quatuoᷓ rectis lineis: q̄dā multilatere q̄ pluribus q̄
quatuoᷓ rectis lineis continenᷓ. ℂ Figuraᷓ trilaterarum: alia eſt
triangulus babens tria latera equalia. Alia triangulus duo babés
equalia latera. Alia triangulus triū inequaliū lateᷓ. Daᷓū iteruᷓ
alia eſt oᷓthogoniū: vnū. ſ. rectū angulum babens. Alia eſt ambli
gonium aliquem obtuſum angulum habens. Alia eſt oᷓigonium:
in qua tres anguli ſunt acuti. ℂ Figurarum autem quadrilaterarū.
Alia eſt quadratū quod é equilateruᷓ atqʒ rectangulū. Alia eſt te/
tragonus longus: que eſt figura rectangula: ſed equilatera non eſt
Alia eſt belmuayn: que eſt equilatera: ſed rectangula non eſt.

about Euclid himself. He seems to have compiled his work on geometry around 300 B.C.E., probably in Alexandria, and possibly with the help of others. It seems to have eclipsed all previous attempts to put geometry on a sound footing, for only the sketchiest hints of earlier works survive, although we have tantalising glimpses of Greek work on geometry extending back as far as 600 B.C.E. On the other hand, what Euclid wrote found favor with the later Christian church, as well as Arab and Islamic writers, and all thirteen of the books that form his *Elements* survive.

Those books are a mixed bunch, so much so that historians of mathematics have been tempted to assign different phases of Greek mathematics to different books. Very roughly, there are four books about lines, triangles, and circles. Then come two books on ratio and proportion, and then three on properties of numbers – these are supposed to describe some of the earliest Greek work in mathematics. In Book 10, the longest and most obscure, Euclid deals with types of lengths involving square roots of square roots, such as $\sqrt{2 + \sqrt{5}}$, which may seem a strange topic for a work on geometry. But in the last three books he discusses three-dimensional geometry, and in particular the five regular solids (sometimes called the Platonic solids, because of the intense interest Plato took in them in his *Timaeus*). The fifth of these had just been discovered in Plato's day, along with the realization that there are indeed exactly five solids with the property that, in each one, all the faces are the same and

Geometrical definitions in the first printed edition of the Greek text of Euclid's *Elements*, published at Basel in 1533.

every vertex looks alike (in other words, each has the same number of edges meeting at it). The sorts of lengths Euclid discussed typically arise at the ratio of a diagonal of a regular solid to an edge.

It is easy to underestimate the value of what Euclid did in writing his *Elements*. Even if one allows that surveyors and architects might need to know their *Elements* very well, what of the rest of us? At the beginning of the twenty-first century, when Euclidean geometry is disappearing from school syllabuses, it can seem that theorems about triangles and circles are trivial, and epitomize the kind of learning only school teachers out of touch with real needs might value. It will turn out that the opinions of school teachers are important for this story – but first to defend Euclid.

One of the greatest uses for mathematics is to be right. Its conclusions are thought to be arrived at with sufficient clarity to be compelling. Relying on nothing more than elementary Euclidean geometry, tunnelers from Samos in 530 B.C.E. and tunnelers today dig simultaneously from each end, confident that they will meet in the middle. Whenever a science or a technology invokes mathematics, it requires that the mathematics be valid, which means that its arguments are correctly proved. The greatest errors that can be made in setting up any mathematical enterprise are of three kinds. Specific arguments may just turn out to be incorrect. When this happens it may turn out that the terms were not adequately defined and were misleading, so the second type of error is in misunderstanding what the mathematical terms mean. The third error is to assume what you want to prove. This is easy to avoid in short pieces of work, but in a long argument it is a tempting mistake. One argues that conclusion A, which is one you want, is implied by statement B. Now B is true because another statement, C, is. And C is true because D is, and so on, until Z is reached, which is true – because A is! Such a fatal error is called a vicious circle. One of Euclid's finest achievements was in sorting out a set of initial assumptions from which all the conclusions he wanted in elementary geometry could be derived, without there being any vicious circles. He also did well at providing valid proofs on his own terms.

It follows that Euclid's *Elements* is interesting quite independent of its subject matter. It presents a set of logical deductions which are a model of how reasoning can be done. In fact, precisely because his assumptions (some of which we shall look at shortly) are so easy to accept and so bland, the book is a better model than one based on religious, legal, or moral ideas could ever have been. Teachers of any persuasion could advocate the close study of the *Elements* as a preliminary to whatever more important, but more culturally dependent, topics were to come later in their students' education.

Nonetheless, Euclid's *Elements* is about geometry. It concerns figures in the plane and in three-dimensional space, and as such it is about space. The easier of these to understand is the plane, so when-

Tetraedron Planum Vacuum

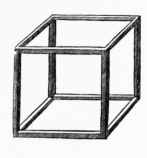

Hexaedron. Planum. vacuum.

Octaedron Planum Vacuum

Dodecaedron Planum Vacuum

Icosaedron Planum Vacuum

The Platonic solids,
as depicted in Luca Pacioli's
Divina Proportione
(Venice, 1509).

ever possible people concentrate on it and hope that they can solve their mathematical problem that way. We shall do the same. What do we learn about the plane by reading Euclid's *Elements*? We learn that lines may be extended indefinitely – there is no edge to the plane; that figures may be scaled up and down arbitrarily in size; and that all regions of the plane are alike – what can be done in one place can be done anywhere. The plane, it seems, is flat and featureless; it has no properties itself that could affect the objects in it. And the same is true of Euclidean space.

This is what every schoolchild learns about physical space, so it is interesting to see that this was not the educated Greek view. The well-educated Greek may well have subscribed both to Euclid's *Elements* as an account of mathematical space and to Aristotle's account of finite physical space. There is no contradiction there, only the apparent fact that a certain mathematical abstraction, Euclidean space, is not actual physical space. But it alerts us to the fact that the truth of Euclid's *Elements* is not as easy to understand as it might be. We shall return to this point later on.

Euclid did not begin with an agenda and then contrive hypotheses to suit it; or, if he did, he had enough rhetorical sense to keep the agenda hidden. The *Elements* does not begin with claims about the nature of space. They begin with a few definitions of fundamental objects, such as points, lines and planes, and even these may be later interpolations. Euclid himself may have begun with definitions of objects that

BOLYAI'S APPENDIX

are less well known and require definition before communication can properly take place. These are the definition of an angle, specifying a term for the angles between straight lines, as opposed to curved arcs, and defining various types of angle according to their size and various types of straight-sided figures according to the number of sides they have.

He then makes a number of assumptions. Some are called Common Notions, which are rules of reasoning applicable to any kind of quantity, such as: "Things which are equal to the same thing are equal to each other," and "If equals be added to equals, the wholes are equal." His Postulates are more explicitly geometrical. These allow us to assert that any two points can be joined by a line, and any line can be extended. It emerges that this extension can be of any size. That space is the same at each point follows from the Postulate that all right angles are equal to each other, and that circles can be constructed with any given center and radius. Prominent among the Postulates, and the one with which, we shall see, Bolyai was so successfully to disagree, is the parallel postulate. In its original form it says:

"That, if a straight line falling on two straight lines make the interior angles on the same side less than two right angles, the two straight lines, if produced indefinitely, meet on that side on which are the angles less than two right angles."

To see what this means, consider Figure 1. Two lines, *l* and *n*, are crossed by a third, *m*, and the an-

Δύο γωνίας τὰς ὑπὸ ᾱβγ βγᾱ, δυσὶ ταῖς ὑπὸ δεζ εζδ, ἴσας ἔχοντα ἑκατέραν ἑκατέρα
τὼ μὲ ὑπὸ ᾱβγ, τῇ ὑπὸ δεζ, τὼ δὲ ὑπὸ βγᾱ, τῇ ὑπὸ εζδ. ἐχέτω δὲ καὶ μίαν πλευρὰ μι
μιᾷ πλευρᾷ ἴσην πρότερον, τὼ πρὸς ταῖς ἴσαις γωνίαις τῇ βγ τῇ εζ, λέγω ὅτι καὶ τὰς λα
τὰς πλευρὰς, ταῖς λοιπαῖς πλευραῖς ἴσας ἕξει ἑκατέραν ἑκατέρα, τὼ μὲ ᾱβ τῇ δε, τὼ
δὲ ᾱγ, τῇ δζ, καὶ τῇ λοιπὴν γωνίαν, τῇ λοιπῇ γωνίᾳ, τὼ ὑπὸ βᾱγ, τῇ ὑπὸ εδζ. εἰ γὰρ ἄ
σος ἐστὶ ἡ ᾱβ, τῇ δε, μία αὐ τῶ μείζων ἐστι. ἔστω μείζω, ἡ ᾱβ. καὶ κείσθω τῇ δε ἴση ἡ βθ
ἐπιζεύχθω ἡ θγ. ἐπεὶ ὀν ἴση ἐστὶ ἡ μὲ βθ, τῇ δε, ἡ δὲ βγ, τῇ εζ. δύο δὴ αἱ βθ βγ, δυο
ταῖς δε εζ, ἴσαι εἰσὶ ἑκατέρα ἑκατέρα, καὶ γωνία ἡ ὑπὸ θβγ, γωνίᾳ τῇ ὑπὸ δεζ, ἴση ἐστὶ
Βάσις ἄρα ἡ θγ, Βάσει τῇ δζ ἴση ἐστὶ. καὶ τὸ θβγ τρίγωνον, τῷ δεζ τριγώνῳ ἴσον ἐστα, ὡ
αἱ λοιπαὶ γωνίαι, ταῖς λοιπαῖς γωνίαις ἴσαι ἔσονται ἑκατέρα ἑκατέρα, ὑφ᾽ ἃς αἱ ἴσαι πλε
ραὶ ὑποτείνουσιν. ἴση ἄρα ἡ ὑπὸ θγβ γωνία, τῇ ὑπὸ δζε. ἀλλὰ ἡ ὑπὸ δζε, τῇ ὑπὸ βγᾱ ὑπο
κεῖται ἴση. καὶ ἡ ὑπὸ βγθ ἄρα, τῇ ὑπὸ βγᾱ ἴση δὴ, ἡ ἐλάσσων τῇ μείζονι, ὅπερ ἀδύνατον
οὐκ ἄρα ἄνισός ἐστιν ἡ ᾱβ, τῇ δε, ἴση ἄρα. ἔστι δὲ καὶ ἡ βγ τῇ εζ ἴση, δύο δὴ αἱ ᾱβ, βγ, δύο
ταῖς δε εζ, ἴσαι εἰσὶν ἑκατέρα ἑκατέρα, καὶ γωνία ἡ ὑπὸ ᾱβγ, γωνίᾳ τῇ ὑπὸ δεζ ἴ
ἴση, Βάσις ἄρα ἡ ᾱγ, Βάσει τῇ δζ, ἴση δὴ, καὶ λοιπὴ γωνία ἡ ὑπὸ βᾱγ, λοιπῇ γωνία τῇ ὑπ
εδζ ἴση δὴ. ἀλλὰ διὰ πάλιν ἐσωσιν αἱ ὑπὸ τὰς ἴσας γω-
νίας πλευραὶ ὑποτείνουσαι ἴσας, ὡς ἡ ᾱβ, τῇ δε, λέγω πά-
λιμ, ὅτι καὶ αἱ λοιπαὶ πλευραὶ, ταῖς λοιπαῖς πλευραῖς ἴσαι
ἔσονται, ἡ μὲ ᾱγ, τῇ δζ, ἡ δὲ βγ τῇ εζ, ἔτι ἡ λοιπὴ γωνία ἡ ὑπὸ βᾱγ, λοιπῇ τῇ ὑπὸ εδζ ἴση δὴ. εἰ γὰρ ἄνισός
ἐστιν ἡ βγ τῇ εζ, μία αὐ τῶ μείζων ἐστὶ. ἔστω εἰ δυνατὸν μείζων,
ἡ βγ, καὶ κείσθω τῇ εζ ἴση ἡ βθ. ἐπεζεύχθω αἱ θᾱ. καὶ ἐπεὶ
ἴση ἐστὶ ἡ μὲ βθ, τῇ εζ, ἡ ᾱβ τῇ δε, δύο δὴ αἱ ᾱβ, βθ, δύο
δυσὶ ταῖς δε εζ, ἴσαι εἰσὶν ἑκατέρα ἑκατέρα, καὶ γωνίας
ἴσας περιέχουσι, Βάσις ἄρα ἡ ᾱθ, Βάσει τῇ δζ ἴση δὴ, καὶ τὸ ᾱβθ τρίγωνον, τῷ δεζ τριγώνῳ ἴσον δὴ, ὡς αἱ λοιπαὶ γωνίαι, ταῖς λοιπαῖς γωνίαι ἴσονται ἑκατέρα ἑκατέρα, ὑφ᾽ ἃς αἱ ἴσαι
πλευραὶ ὑποτείνουσιν. ἴση ἄρα ἐστὶ ἡ ὑπὸ βθᾱ γωνία, τῇ ὑπὸ εζδ, ἀλλὰ ἡ ὑπὸ εζδ, τῇ ὑπὸ
βγᾱ γωνίᾳ, ἐστὶν ἴση. καὶ ἡ ὑπὸ βθᾱ ἄρα, τῇ ὑπὸ βγᾱ ἴση ἐστι, τριγώνου δὴ τοῦ ᾱθγ, ἡ ἐκ
τὸς γωνία ἡ ὑπὸ θᾱ ἴση ἐστὶ τῇ ὑπὸ θγᾱ, καὶ εἰ τῷ ἐναντίῳ τῇ ὑπὸ θγᾱ, ὅπερ ἀδύνατον. οὐκ ἄρα
ἄνισός ἐστιν ἡ βγ, τῇ εζ, ἴση ἄρα. ἐστὶ δὲ καὶ ἡ ᾱβ, τῇ δε ἴση, δύο δὴ αἱ ᾱβ, βγ, δύσι ταῖς δ
εζ, ἴσαι εἰσὶν ἑκατέρα ἑκατέρα, καὶ γωνίας ἴσας περιέχουσι. Βάσις ἄρα ἡ ᾱγ, Βάσει τῇ δζ ἴση
δὴ, καὶ τὸ ᾱβγ τρίγωνον, τῷ δεζ τριγώνῳ ἴσον δὴ, καὶ λοιπὴ γωνία ἡ ὑπὸ βᾱγ τῇ λοιπῇ
γωνίᾳ τῇ ὑπὸ εδζ ἴση ἐστὶ. ἐὰν ἄρα δύο τρίγωνα τὰς δύο γωνίας ταῖς δυσὶ γωνίαις ἴσας
ἔχη ἑκατέραν ἑκατέρα, καὶ μίαν πλευρὰν μιᾷ πλευρᾷ ἴσην ἔχη, ἤτοι τῇ πρὸς ταῖς ἴσαις γωνίαις ἡ τῇ ὑποτεινούσῃ ὑπὸ μίαν τῶν ἴσων γωνιῶν, καὶ τὰς λοιπὰς πλευρὰς, ταῖς λοιπαῖς
πλευραῖς ἴσας ἕξει. καὶ τὴν λοιπὴν γωνίαν τῇ λοιπῇ. ὅπερ ἐδει δεῖξαι.

ΕΑΝ εἰς δύο εὐθείας εὐθεῖα ἐμπίπτησα τὰς ἐν ἀλλὰξ γωνίας ἴσας ἀλλήλαις ποιῇ, παράλληλοι ἔσονται ἀλλήλαις αἱ εὐθεῖαι. Εἰς γὰρ δύο εὐθείας
τὰς ᾱβ γδ, δύο εὐθεῖα ἐμπίπτησα ἡ εζ, τὰς ἐν ἀλλὰξ γωνίας τὰς ὑπὸ
ᾱεζ εζδ, ἴσας ἀλλήλαις ποιείτω, λέγω ὅτι παράλληλός ἐστιν ἡ ᾱβ,
τῇ γδ εὐθεία. εἰ γὰρ μὴ, ἐκβαλλόμεναι αἱ ᾱβ γδ συμπεσοῦνται
ἤτοι ἐπὶ τὰ β δ μέρη, ἡ ἐπὶ τὰ ᾱγ. ἐκβεβλήσθωσαν καὶ συμπι-
πτέτωσαν ἐπὶ τὰ β δ μέρη κατὰ τὸ θ. τριγώνου δὴ τοῦ θεζ ἡ ἐκτὸς
γωνία ἡ ὑπὸ ᾱεζ, † μείζων ἐστὶ τῆς ἐντὸς καὶ ἀπεναντίου γωνίας τῆς
ὑπὸ εζθ, ἀλλὰ καὶ ἴση, ὅπερ ἐστὶν ἀδύνατον. οὐκ ἄρα αἱ ᾱβ γδ ἐκβαλλόμεναι συμπεσοῦνται, ἐπὶ τὰ β δ μέρη. ὁμοίως δὴ δειχθή-
σεται, ὅτι οὐδὲ ἐπὶ τὰ ᾱγ, αἱ δὲ εἰς μηδέτερα τὰ μέρη συμπίπτου-
σαι, παράλληλοί εἰσι, παράλληλος ἄρα ἐστὶν ἡ ᾱβ, τῇ γδ, ἐὰν ἄρα εἰς δύο εὐθείας εὐθεῖα ἐμπι
πτέτωσα τὰς

aliud exemplar
sic,io τὸ τὸ ισ-
τα καὶ αἱ γραμ-
τιον τ᾽ὖ ὖσι ἰ-
ἐπιζ ἀσθ᾽σωρ.
habet utrúque
recte.

ἥπτεσαι τὰς ὀναλλὰξ γωνίας ἴσας ἀλλήλαις ποιῇ, πὸ ἄλληλοι ἴσονται αἱ εὐθεῖαι. ὅπρ ἔδ'ει δεῖξαι.

ΕΑΝ εἰς δύο εὐθείας εὐθεῖα ἐμπίπτᾳ, τὴν ἐκτὸς γωνίαν, κ)
τῇ ἐντὸς κỳ ἀπεναντίον κỳ ἐπὶ τὰ αὐτὰ μόρῃ ἴσην ποιῇ, κỳ τὰς ἐντὸς Ͼ
ἐπὶ τὰ αὐτὰ μόρῃ δυσὶν ὀρθαῖς ἴσας ποιῇ, πὸ ἄλληλοι ἴσονται ἀλλήλαις
αἱ εὐθεῖαι. Εἰς γὰρ δύο εὐθείας τὰς αβ γδ, εὐθεῖα ἐμπιπτέτω ἡ εζ, τὴν ἐκτὸς γωνίαν Ͷ
ἴσον, τὴν ἐντὸς Ͼ τὴ ἀπεναντίον γωνίαν τῇ ὑπὸ ἡθδ, ἴσην ποιείτω, ἢ τὰς ἐντὸς κỳ ἐπὶ τὰ
αὐτὰ μόρῃ τὰς ὑπὸ βηδ λέγω ὅτι παράλληλός ἐδ'ιν ἡ αβ, τῇ γδ.
ἐπεὶ γὰρ ἴση ὀδὶν ἡ ὑπὸ εηβ, τῇ ὑπὸ ηθδ, ἀλλὰ ἡ ὑπὸ εηβ, τῇ ὑπὸ αηθ ἴση, κ) ἡ ὑπὸ αηθ
ἄρα, τῇ ὑπὸ ηθδ ἴση ἐδ'ιν. κỳ εἰσιν ἐναλλάξ. ἀλλ' ἄρα ἐδ'ιν ἡ αβ. τῇ γδ. πάλιν ἐπεὶ
αἱ ὑπὸ βηθ κỳ ηθδ, δυσὶν ὀρθαῖς ἴσαι εἰσιν, εἰσὶ δὲ κỳ αἱ ὑπὸ αηθ Ͼ βηθ δυσὶν ὀρθαῖς ἴσαι, αἱ
ἄρα ὑπὸ αηθ βηθ, ταῖς ὑπὸ βηθ ηθδ, ἴσαι εἰσι, κοινὴ
ἀφῃρήσθω ἡ ὑπὸ θηβ, λοιπὴ ἄρα, ἡ ὑπὸ αηθ, λοιπῇ τῇ ὑπὸ
ηθδ ἴση ἐδ'ιν. Ͼ εἰσὶν ἐναλλάξ. ἀλλ' ἄρα ἐδ'ιν ἡ αβ
τῇ γδ. κỳ ἄρα εἰς δύο εὐθείας εὐθεῖα ἐμπίπτουσα, τὴν ἐκ
τὸς γωνίαν τῇ ἐντὸς κỳ ἀπεναντίον Ͼ ἐπὶ τὰ αὐτὰ μόρῃ
ἴσην ποιῇ, τὰς ἐντὸς κỳ ἐπὶ τὰ αὐτὰ μόρῃ δυσὶν ὀρθαῖς
ἴσας, παράλληλοι ἴσονται αἱ εὐθεῖαι. ὅπρ ἔδ'ει δεῖξαι.

Η εἰς τὰς παραλλήλους εὐθείας εὐθεῖα ἐμπίπτᾳ, τάς τε ὀναλλὰξ γωνίας ἴσας ἀλλήλαις ποιεῖ, κỳ τὴν ἐκτὸς, τῇ ἐντὸς κỳ ἀπεναντίον, κỳ ἐπὶ
τὰ αὐτὰ μόρῃ, ἴσην, Ͼ τὰς ἐντὸς κỳ ἐπὶ τὰ αὐτὰ μόρῃ δυσὶν ὀρθαῖς ἴσας.

Εἰς γὰρ παραλλήλους εὐθείας τὰς αβ γδ, εὐθεῖα ἐμπιπτέτω ἡ εζ, λέγω ὅτι τὴν τε ὀναλλὰξ γωνίας τὰς ὑπὸ αηθ, ἴσας ποιεῖ, κỳ τὴν ἐκτὸς γωνίαν τὴν ὑπὸ εηβ, τῇ ἐντὸς κỳ
ἀπεναντίον κỳ ἐπὶ τὰ αὐτὰ μόρῃ τῇ ὑπὸ ηθδ ἴσην, κỳ τὰς ἐντὸς Ͼ ἐπὶ τὰ αὐτὰ μόρῃ τὰς
ὑπὸ βηθ ηθδ, δυσὶν ὀρθαῖς ἴσας. εἰ γὰρ ἄνισός ἐδ'ιν ἡ ὑπὸ αηθ τῇ ὑπὸ ηθδ μεῖζων
ἐδ'ιν. ἔστω μείζων, ἡ ὑπὸ αηθ, κỳ κοινὴ προσκείσθω
ἡ ὑπὸ βηθ, αἱ ἄρα ὑπὸ αηθ βηθ, τῆς ὑπὸ βηθ ηθδ μείζονές εἰσιν. ἀλλὰ κ) αἱ ὑπὸ αηθ
βηθ, δυσὶν ὀρθαῖς ἴσαι εἰσίν· αἱ ἄρα ὑπὸ βηθ ηθδ, δύο ὀρθῶν
ἐλάσσονές εἰσιν. αἱ δὲ ἀπ' ἐλασσόνων ἢ δύο ὀρθῶν ἐκβαλλό
μεναι εἰς ἄπειρον, συμπίπτουσιν. αἱ ἄρα αβ γδ, ἐκβαλλόμεναι
εἰς ἄπειρον, συμπεσοῦνται, οὐ συμπίπτουσι δὲ, διὰ τὸ παραλλήλους αὐτὰς ὑποκεῖσθαι. οὐκ ἄρα ἄνισός ἐδ'ιν ἡ ὑπὸ αηθ, τῇ ὑπὸ
ηθδ ἄρα, τῇ ὑπὸ ηθδ ἴση ἐδ'ιν, ἀλλὰ ἡ ὑπὸ αηθ τῇ ὑπὸ εηβ ἴση·
κ) ἡ ὑπὸ εηβ, τῇ ὑπὸ ηθδ ἴση. κοινὴ προσκείσθω, ἡ ὑπὸ βηθ,
αἱ ἄρα ὑπὸ εηβ βηθ, ταῖς ὑπὸ βηθ ηθδ ἴσαι εἰσι· ἀλλὰ αἱ
ὑπὸ εηβ βηθ, δυσὶν ὀρθαῖς ἴσαι εἰσι, κỳ αἱ ὑπὸ βηθ ηθδ ἄρα, δυσὶν ὀρθαῖς ἴσαι εἰσίν. κ)
ἄρα εἰς τὰς παραλλήλους εὐθείας εὐθεῖα ἐμπίπτουσα, τὰς τε ὀναλλὰξ γωνίας ἴσας ἀλλήλαις
ποιεῖ, κỳ τὴν ἐκτὸς, τῇ ἐντὸς κỳ ἀπεναντίον, κỳ ἐπὶ τὰ αὐτὰ μόρῃ ἴσην, κỳ τὰς ἐντὸς κỳ
ἐπὶ τὰ αὐτὰ μόρῃ, δυσὶν ὀρθαῖς ἴσας. ὅπρ ἔδ'ει δεῖξαι.

ΑΙ τῇ αὐτῇ εὐθείᾳ παράλληλοι, κỳ ἀλλήλαις εἰσὶ παράλληλοι.

Ἔστω ἑκατέρα τῶν αβ γδ, τῇ εζ δὲ παράλληλ⊙.
λέγω ὅτι κ) ἡ αβ, τῇ γδ ἐδ'ι παράλληλ⊙. ἐμπιπτέτω
γὰρ εἰς αὐτὰς εὐθεῖα ἡ ηκ. κỳ ἐπεὶ εἰς παραλλήλους εὐθείας τὰς
αβ εζ, εὐθεῖα ἐμπέπτωκεν ἡ ηκ, ἴση ἄρα ἡ ὑπὸ αηθ, τῇ ὑπὸ
ηθζ. πάλιν ἐπεὶ εἰς τὰς παραλλήλους εὐθείας τὰς εζ γδ, εὐθεῖα ἐμπέπτωκεν ἡ ηκ, ἴση ἐδ'ιν ἡ ὑπὸ ηθζ, τῇ ὑπὸ ηκδ, ἐδείχ-
θη δὲ ἡ ὑπὸ ηηκ, τῇ ὑπὸ ηθζ ἴση, κ) ἡ ὑπὸ ηηκ ἄρα, τῇ
ὑπὸ ηκδ ἴση, κỳ εἰσὶν ὀναλλάξ, παράλληλ⊙ ἄρα ἐδ'ιν ἡ

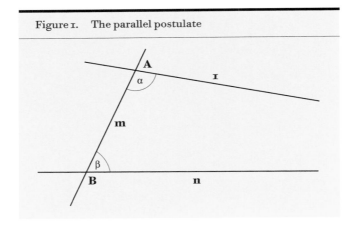

Figure 1. The parallel postulate

gles made at the points *A* and *B* where they cross are α and β, and we are to suppose that the angle sum α + β is less than two right angles. The postulate says that if we extend the lines far enough on the side where the angles are (to the right in this case) the lines *l* and *n* will eventually meet.

Two things are excluded by this assumption. It is not the case that if the lines are extended backwards, to the left, they will then meet. Nor is it the case that the lines never meet, in either direction. Later, in Proposition 4 of Book 1, Euclid assumes that two distinct lines cannot meet in two points and so enclose a space. This appears to have been assumed without comment by Euclid and only spelled out by a later commentator, perhaps the Arabic writer al-Naiziri, before becoming part of the standard texts of the *Elements*.

BOLYAI'S APPENDIX

The Parallel Postulate

THE PARALLEL POSTULATE is an assumption, and a remarkable one. It is easy to imagine figures where the angles α and β sum to something so close to two right angles that the meeting point is very far away, let us say in a distant galaxy. On what grounds then should we assume it? And it is a property of straight lines, for Greek geometers knew very well that the hyperbola gets indefinitely close to its asymptote but never meets it (Figure 2). So what exactly is it about straight lines that justifies the parallel postulate?

The Greeks had no good answers to either question, although we know from fragments of Aristotle's writings that they discussed them – giving rise to a modern scholarly article on non-Euclidean geometry before Euclid! But they had a tacit answer: it was

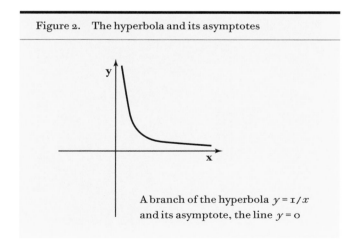

Figure 2.   The hyperbola and its asymptotes

A branch of the hyperbola $y = 1/x$
and its asymptote, the line $y = 0$

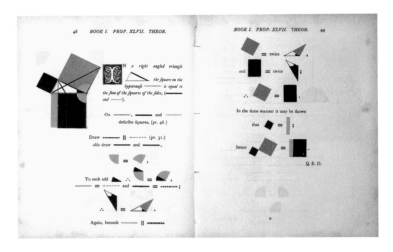

The Pythagorean theorem, from Oliver Byrne's 1847 edition of
the *Elements*, in which all equations were replaced
with colored diagrams.

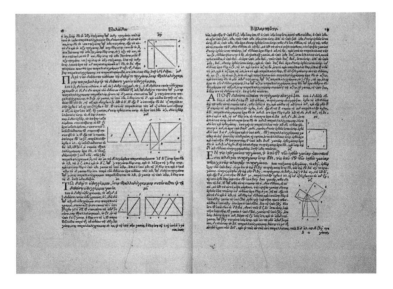

A more orthodox version of the Pythagorean theorem,
from the 1533 edition of the *Elements*.

Box 1. The angle sum of a triangle in Euclidean geometry

We are given triangle ABC with angles α, β, and γ respectively, and to show that α + β + γ equals two right angles. This is done by extending the side AC to a point E and drawing the line CD parallel to AB. The angles at the point C sum to two right angles. But because the lines CD and AB are parallel, the angle BCD equals the angle ABC (which is β) and the angle BAC (which is α) equals the angle DCE. So the sum of the angles, α + β + γ, equals two right angles.

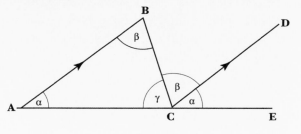

---

useful. Indeed, without the parallel postulate, much of Euclid's *Elements* cannot be written.

For example, the proof of the famous Theorem of Pythagoras, which states that in any right-angled triangle the sum of the squares on the smaller sides is equal to the square on the largest side, uses the parallel postulate. One can show using the parallel postulate that in any triangle the three angles sum to exactly two right angles (as Euclid does in Book 1, Proposition 32; see also Box 1). One can show that two parallel lines, like railway lines, are everywhere the same distance apart. The theory of ratio and proportion, and the construction of triangles having the same shape but different sizes, uses the parallel postulate; so it is the parallel postulate that allows us to make accurate scale copies of figures. The parallel postulate is so useful because it implies that given a line *l* and point *P* not on the line, there is a unique line through *P* that never meets *l*. This line is called the parallel to *l* though *P*, and its uniqueness is what allows one to find equal angles in figures involving parallel lines and deduce the results just quoted.

On the other hand, however useful, even essential the parallel postulate might be, it is awkward to assume it, because it cannot be said to be a simple truth of experience. Other Greek writers confronted the problem and attempted to deduce the parallel postulate from the other assumptions and definitions of the *Elements*. In this they failed (some of their arguments are discussed in Gray [1989]), and later in this book we shall see why. But Euclid did not mar his *Elements* with a fallacious proof, which is one more reason why it was able to grow in stature with its transmission to the West.

Euclid's *Elements* came to be regarded as the epitome of reasoning, but they were not all of mathematics, nor was Euclidean geometry the only way to do mathematics. The *Elements* was not even all of Greek mathematics. There was a remarkable study of conic sections (the ellipse, parabola, and hyperbola) made by Apollonius, and a number of special problems were discussed. One of these should be men-

BOLYAI'S APPENDIX

tioned here: it was required to show that, using only circles and straight lines, a square can be constructed equal in area to a given circle (or else that it cannot be done). The problem of squaring the circle, as it was called, rapidly became a synonym for impossibility. The Athenian playwright Aristophanes could raise a laugh by alluding to it in his play *The Birds* – which says something about the erudition of his audience and the fame of the problem.

What Western writers objected to after, say, 1600, was that it was very difficult to do mathematics in the style of the Greeks. In particular, with the spread of the mathematical ideas and methods put forward by René Descartes in his *La Géométrie* (1644), it became more and more usual to do geometry with a mixture of largely algebraic methods, which evolved into what today is called Cartesian coordinate geometry. What re-established Euclidean geometry was a work of theoretical astronomy, one that gave it a central role in explaining, and not merely describing, the motion of the planets. This was the remarkable *Principia Mathematica* of Sir Isaac Newton, published for the first time in 1687. With it we find a persuasive mechanics, linked to an account of physical space that is through-and-through Euclidean.

Newton decided not to define time, space, place, and motion "as being well known to all," which might have been his way of evading tasks notoriously difficult. As the familiar saying has it "I know what time is, until I think about it." But to remove "certain prej-

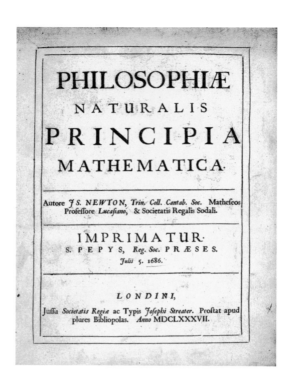

The title page of the first edition of Sir Isaac Newton's
Principia Mathematica.

udices" concerning absolute and relative time, and
absolute and relative space, he added:

> Absolute space, of its own nature without refer-
> ence to anything external, always remains ho-
> mogenous and immovable. Relative space is any
> movable measure or dimension of this absolute
> space; such a measure or dimension is determined
> by our senses from the situation of the space with

BOLYAI'S APPENDIX

Sir Isaac Newton

respect to bodies and is popularly used for im-
movable space, . . . place is a part of space that a
body occupies.[1]

The physical space that Newton described oblit-
erated any distinction between the space beneath the
moon and the space in which the planets moved, a
distinction already firmly eroded by the work of Coper-
nicus, Kepler, and Galileo; and by clear implication
Newton extended this unity to the stars. To this one
physical space he brought novel Euclidean theorems.
The *Principia* is written in the language of Euclidean
geometry, applied simultaneously to finite and infini-
tesimal regions with extraordinary ingenuity. The
force of gravity, which Newton carefully said he had
not explained, was a tremendous novelty.[2] It was quite
unclear how it could act over such large distances as
those in the solar system, but after Newton's prodi-
gious mathematical efforts, directed as much at re-
futing Cartesian ideas as shoring up his own, there
was to be little doubt that a force acted between any
two bodies that was proportional to their masses and
fell off inversely as the square of the distance between
them. The identification of physical space with the
space of Euclidean geometry was complete.

The geometer Clairaut, who was influential in
winning French support for the inverse square law of
gravitation, and successfully predicted the date of
the return of Halley's comet, was a Newtonian in his
mathematics and a Cartesian in his metaphysics. He

René Descartes

grounded geometry not on the postulates of Euclid but on the capacity of the mind to form clear and distinct ideas. In the preface to his *Eléments de Géométrie* (1741) he criticized the usual textbooks on geometry, which "always begin with a great number of definitions, postulates, axioms and preliminary principles which seem to promise the reader nothing but dry-

Alexis-Claude Clairaut

ness." Nor, he went on, does it help to back up each proposition with some indication of its practical use; this "proves the utility of geometry without making it easier to understand." He therefore proposed to present geometry somewhat in the fashion of its his-

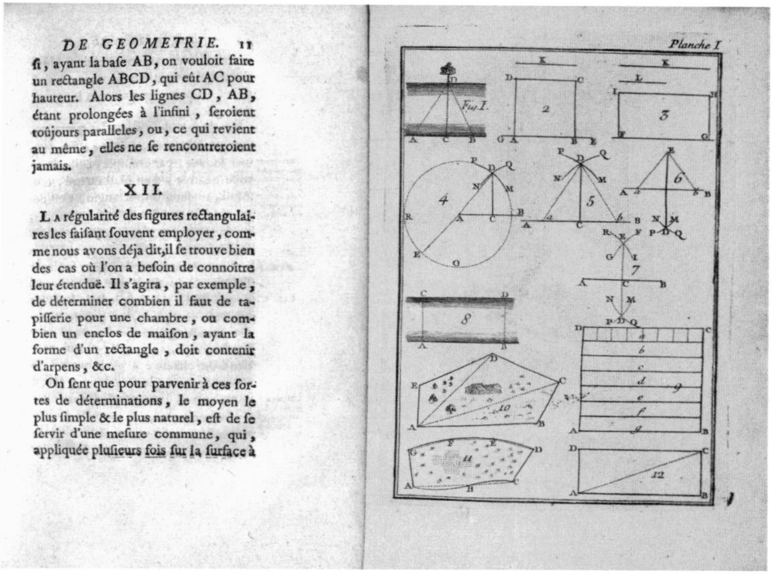

DE GEOMETRIE. 11

si, ayant la base AB, on vouloit faire un rectangle ABCD, qui eût AC pour hauteur. Alors les lignes CD, AB, étant prolongées à l'infini, seroient toûjours paralleles, ou, ce qui revient au même, elles ne se rencontreroient jamais.

XII.

La régularité des figures rectangulaires les faisant souvent employer, comme nous avons déja dit, il se trouve bien des cas où l'on a besoin de connoître leur étenduë. Il s'agira, par exemple, de déterminer combien il faut de tapisserie pour une chambre, ou combien un enclos de maison, ayant la forme d'un rectangle, doit contenir d'arpens, &c.

On sent que pour parvenir à ces sortes de déterminations, le moyen le plus simple & le plus naturel, est de se servir d'une mesure commune, qui, appliquée plusieurs fois sur la surface à

Paragraphs eleven and twelve of Clairaut's *Éléments de Géométrie* (Paris, 1741).

torical order, to keep its practical use in surveying continually in view, and to enable his readers to solve problems. Accordingly, many propositions of Euclid's *Elements* would not appear in his: "I can be criticized, perhaps, at some points in these *Éléments*, for relying too much on the evidence of the eyes and not attaching enough importance to the rigorous exactitude of proofs." But, he said, times have changed and it is not right to obscure the truth and fill the reader with distaste. Clairaut himself had no earthly or celestial reason for doubting the truth of Euclidean geometry.

The resulting book has very few definitions and

much common sense. It is divided into paragraphs, of which number eleven begins "Parallels are straight lines which are equidistant the one from the other" and immediately discusses the construction of canals and streets. This blunt, no-nonsense approach starts with the most useful property of parallels, but not, as we shall see, with the simplest. Euclid had taken the view that a given pair of lines either meet or they do not, and if they do not then certain other properties follow. He merely assumed that it was always possible to say when they do, or do not, meet. Since Euclid and Clairaut agreed that there are parallel lines, and such lines are everywhere equidistant, it might seem that it does not matter that their foundations of geometry are different. But when results about parallels were called into question (by Bolyai and Nikolai Ivanovich Lobachevskii) the implications of that challenge were much clearer for Euclidean-style texts than the practical texts *à la* Clairaut.

A generation later we find both the French Encyclopedistes and German philosopher Immanuel Kant in agreement over the Newtonian synthesis of geometric and physical space. Jean le Rond d'Alembert hailed the *Principia* as "the most extensive, the most admirable, and the happiest application of geometry to physics which has ever been made," and he too defined parallel lines as everywhere equidistant. Kant, in his *Critique of Pure Reason* (1781) wished to raise metaphysics to the level of science (Kant [1781], 23). His aim was to show how such knowledge was possible, and he valued it very highly. So he wrote: "Pure mathematics is a brilliant example of [*a priori* synthetic] knowledge, especially as regards space and its relations." (Kant [1781], 80). He argued for a philosophy that could explain how geometry is a science which determines the properties of space *a priori*, and of course that knowledge was encapsulated in Euclidean geometry (Kant [1781], 70). This is clearest at the end of the book, when he discussed how a mathematician can know things about space that a philosopher cannot. The answer is that the mathematician makes constructions, and so can deal with quantities. The example Kant gave is precisely the proof of Proposition 32, Book 1 of Euclid's *Elements*, where the parallel to a side is drawn. Quite deliberately, Kant chose an example which establishes that space is validly described by Euclidean geometry.

To be sure, Kant's idea of space was a complicated and deep one, but he claimed that our certain knowledge of it is the knowledge taught by Euclid and his successors, and triumphantly put to use in the study of science. He went further, and identified all of mathematics, geometry and algebra, with the study of quantity. Quantities come in various kinds, notably the discrete and the continuous, the quantities which can be counted and those which must be measured. But the paradigmatic quality of the knowledge was supreme in each case.

The first signs of a shift came with the French Revolution, and the creation of a whole new system of

The *École polytechnique* around 1910,
in a photograph by Eugène Atget.

higher education. This was focused on the newly
founded *École Polytechnique*, which was set up as a
military school to train students for two years after
which they would go on to one of four engineering
schools and then to civil or military careers. The institution was a great success, competition to enter it
was intense, and academic standards rose across
France. Paradoxically, it did not give its charges a very
practical education, but increasingly emphasized the
abstract side of mathematics. There are a number of
reasons for this. Its influence upstream, so to speak,
on the higher reaches of school education, was to pro-
mote mathematics. This was interpreted as a train-
ing for the mind, and as such valuable independent

BOLYAI'S APPENDIX

of the precise content. The art of rigorous thinking could be taught through mathematics, even to those with no intention of using it in later life, and the man so educated would be a better citizen. Downstream, it was necessary to train students for a variety of engineering jobs: how much better to give them a general, abstract education, from which the applications can be derived later, than to specialize too early.

These are familiar arguments to this day. They played better in Paris in 1800 because they were advanced by mathematicians to mathematicians. Officially, the control of the *École Polytechnique* rested with the Army, but the teaching was done by mathematicians whose interests were largely pure, and they could influence the syllabus. The founder of the school was a practical man, Gaspard Monge, who had pioneered a method of adapting geometry to use in architecture, but he was gradually pushed aside. Teaching in geometry became the province of Adrien-Marie Legendre, a much more able mathematician. He brought back a more Euclidean approach, and because his books were the ones every aspiring *polytechnicien* had to read, the shift to this formal style was rapid, complete, and lasting (for example, the twelfth edition of his *Éléments de Géométrie* was published in 1823 and reprinted as the fourteenth in 1860).

Legendre's book was not exactly Euclid's. But it does open with many of Euclid's definitions, and it proves many things Clairaut wanted his readers to take for granted. Parallel lines are defined as lines that do not meet (their existence is assumed). Legendre also took a different route through some of Euclid's elementary theorems. First he proved that the sum of the angles in a triangle is always two right angles, then he proved the parallel postulate, and then he proved that parallel lines are everywhere equidistant. For those with a taste for deducing as much as possible from as little as possible, this is all to the good. It is however with the parallel postulate that it all starts to unravel, and our story takes off.

The Blot

THE PARALLEL POSTULATE as Euclid had expressed it had always annoyed mathematicians. Over the centuries Greek, Arab and Islamic, and finally Western writers had tried to replace it with something more intuitive, or to derive it from the other assumptions and postulates alone. A German doctoral thesis by Georg Simon Klügel (a student of Abraham Kästner) written in 1763 recorded twenty-eight attempts on the parallel postulate. The very fact of the procession indicates that something was amiss, else one of these attempts would surely have commanded assent thereafter. If Euclid's *Elements* is a paradigm of logical virtue, then the parallel postulate is a blot (such was the term used by Henry Savile at Oxford in 1621).

Legendre did not raise the question of the parallel postulate because he had any doubts about the an-

Adrien-Marie Legendre

swer, but because it seemed to him unnecessary to assume it when it could be proved. It was already well-known that Euclid's *Elements* without the parallel postulate implies one of only three possibilities.

1. Either the parallel postulate is true, and the angle sum of every triangle is two right angles;

2. Or every pair of lines meet and the angle sum of every triangle is greater than two right angles;

3. Or given a line and point not on it, there are infinitely many lines through the given point that do not meet the given line, and the angle sum of every triangle is greater than two right angles.

It was widely thought desirable that the latter two hypotheses each separately self-destruct. Legendre first took his stand on the observation that the parallel postulate implies that similar, non-congruent

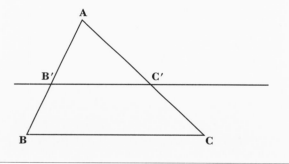
figures exist, and conversely the existence of similar, non-congruent figures implies the parallel postulate. (See Boxes 2.a, 2.b.)

Legendre argued that if the three angles of a triangle determine it up to congruence, then in particular they determine the lengths of the sides, and so the concept of length reduces to that of angle, which he found absurd. In particular it would mean that a unit of length could be determined as a particular function of a standard angle, such as a right angle. His argument is valid as far as it goes, but that is not

Box 2.b. The existence of similar, non-congruent triangles implies the parallel postulate

Let m and m' be two straight lines making angles α and α' as shown with the straight line n, so $\alpha' + \beta'$ is two right angles. We suppose that α' is greater than α, and so $\alpha + \beta'$ is less than two right angles, and seek to show that the lines m and m' meet, thus establishing the parallel postulate.

Consider at each point A'' of the line n drawing a line m'' meeting the line n at an angle of α'. When A'' coincides with A the lines m and m'' certainly meet. Therefore, by continuity, they meet when A'' is very close to A, say at the point B. Call the point where the lines m and m'' meet C. Consider the triangle ABC with base AB and the similar triangle with base AA'. Let its third vertex be the point C'. The angles in these two triangles at the point A are equal, so the second triangle has part of the line m as one of its sides, and the point C' lies on m. The angles in the two triangles at the points B and A' are equal, so the second triangle has part of the line m' as one of its sides, and the point C' lies on m'. It follows that the lines m and m' meet (at the point C') as we wanted to show. This proof was first given by the English mathematician John Wallis in lectures in 1663.

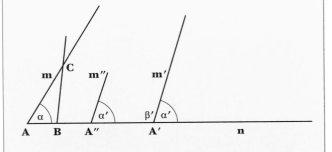

DE

Poſtulato Quinto;

ET

Definitione Quinta

Lib. 6. Euclidis;

DISCEPTATIO GEOMETRICA.

Honoratiſſimus SAVILIUS (à me non ſine Honoris præfatione no-
minandus) in ſuis ad Euclidem *Lecturis*, Duos memorat *Nævos* (ſal-
tem pro talibus habitos) *in pulcherrimo Corpore* Elementorum Euclidis ; quos ſuis recommendat Profeſſoribus eximendos. Nimirum *Quin-
tam Definitionem libri Sexti* ; & *Poſtulatum Quintum* ſeu (ut alii nu-
merant) *Axioma Undecimum*. Quas eſſe *Propoſitiones Demonſtrabiles* contendunt
nonnulli, non gratis *Aſſumendas*.

Verum ego harum neutram culpandam cenſeo.

Def. 5. lib. 6. ſic ſe habet, λέγεται ὅπερ ἐν λόγῳ εὐθεῖαι λέγονται, ὅταν αἱ τῶν λόγων πολυπλάσιαι
πολλαπλασιαζόμεναι, μείζων ποιῇ. (in quibuſdam Codicibus, male habetur πᾶσα pro πᾶ.)
Ubi nihil video difficultatis, niſi quod non ſatis animadverterint ſive Lectores ſive
Interpretes, quid apud *Euclidem* ſignificet πολυπλάσιαι, tum hic, tum in def. 3. lib. 5.
De quibus egimus fuſius (tum alibi, tum) Capp. 19, 20, tractatus de *Algebra*. Ut
non ſit opus hic eadem prolixe repetere.

Summa rei huc redit. In *Rationis* definitione (quæ eſt def. 3. lib. 5.) λόγος ἐστὶ
δύο μεγεθῶν ὁμογενῶν ἡ κατὰ πηλικότητα πρὸς ἄλληλα ποιὰ σχέσις· non exponenda eſt ποιὰ σχέσις
(quod apud plures reperio) *habitudo quædam*, (quaſi foret σχέσις τις,) ſed potius
per πῶς ἔχει, *qualiter ſe habent* : (Et quidem ποιὰ σχέσις non aliter differt à πῶς ἔχει,
quam ut *Nomen Verbale* à *Verbo* ſuo.) Et κατὰ πηλικότητα non tam ſignificat ſecundum quantitatem, quam ſecundum *quantuplicitatem*. Eſt utique *Ratio*, ea *magni-
tudinum homogenearum inter ſe Relatio* (ſeu Habitudo) *qua innuitur quonodo ſe
habet altera ad alteram*, ſecundum *Quantuplicitatem conſiderata*. Hoc eſt, *Quotupla
ſit*, ſeu potius *Quantupla*, altera alterius. Quod ſic intellectum volo : Nimirum,
Prout *Pars Aliquanta* diſtingui ſolet à *parte Aliquota*; ſic ego (quæ ſunt earum
Correlata) *Aliquantuplum* ab *Aliquotuplo* diſtinguo. Ut *Quantuplum* ſit vox
Generalis, cujus *Quotuplum* ſit una Species, cui *Multiplum* reſpondeat.

Eaque *Quantuplicitas* exponi ſolet per (quem dicimus) *Exponentem* Rationis;
hoc eſt per eum numerum (ſeu quod numero eſt homogeneum) unde *Denominari*
ſolet aut *Exponi* Ratio, (ut eſt Rationis *Duple*, numerus 2 ; *Triple*, 3 ; *Sub-
duple*, ½ ; *Subtriple*, ⅓ ; *ſeſquialterius*, 1½ ſeu 3⁄2 : & rationis A ad B, $\frac{A}{B}$) Nempe
Quotiens, qui ex Antecedente per Conſequentem diviſa oritur; quique in Conſequentem ductus Antecedentem facit. Quæ Relatio apud Græcos innui ſolet terminatione ᾱδσιον, apud Latinos terminatione *plum*, apud Nos terminatione —*fold*.
Dicit autem Euclides πολυπλάσια potius quam πολλάσια (hoc eſt, *quantuplum* potius
quam *quotuplum*,) ut vox extendi poſſit ad omne genus Rationes, non minus quam
Multiploriem, aut quorum Ratio poſſit *Veris Numeris* exponi.

Et quidem hæſitaret nemo de ſenſu vocis, ſi diceretur *Ratio*, ea (Magnitudi-
num homogenearum) *Relatio* (ſeu Habitudo) qua innuitur *qualiter ſe habet* al-

Pppp   tera

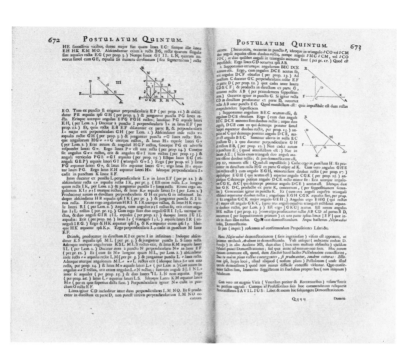

The conclusion of Wallis's proof.

as far as Legendre wanted. Indeed, if the parallel postulate is false then similar figures are congruent and there can be no scale models. This is not something we believe happens in the world. But it is not logically impossible, and the experiences we have that lead us to accept the existence of similar, non-congruent figures are no better or worse than those which incline us to accept the parallel postulate. It would be just as hard to detect minute departures from the similarity of triangles as minute departures from an angle sum of two right angles.

In the next edition Legendre based his argument on the claim that three points which do not lie on a line lie on a circle. The same response was offered: the parallel postulate is as good as this claim, no better, no worse, no more certainly true. From the third to the eighth editions Legendre tried a two-part argument. He showed, correctly, that the angle sum of a triangle cannot exceed two right angles. This can seem surprising (after all, there is a perfectly good geometry on the surface of a sphere, in which the role of straight lines is played by the great circles). Legendre certainly knew this geometry, and in it the angle sum of a triangle exceeds two right angles, by an amount proportional to the area of the triangle. But in this geometry, two lines (for example, two circles of longitude) can enclose a surface, and that was taken to rule it out as a possible geometry of space. More to the point, in this geometry there is a maximum length to a line, given by the circumference of the sphere, and Euclid had assumed that every line segment can be indefinitely extended. So spherical geometry contradicts not just the parallel postulate but two of Euclid's assumptions.

Then Legendre tried to show that the angle sum of a triangle cannot be less than two right angles, and here he ran into difficulties. His argument was to assume that the angle sum of any triangle is less than two right angles, and to deduce a contradiction.

Legendre argued that his construction (See Box 3) gave him a new triangle with an angular defect (the

BOLYAI'S APPENDIX

Start with triangle I, with vertices A, B, and C, and
assume that its angle sum is less than two right angles.
Call its angular defect the amount $2R - (\alpha + \beta + \gamma)$. Rotate
the triangle ABC about the mid-point of BC so that
triangle I goes to triangle II and the point A goes to A'.
Draw any line through A' that meets the lines AB and AC,
say at the points B' and C' respectively. Two new
triangles are obtained, III and IV. Legendre claimed that
the angular defect of the large triangle $AB'C'$ is more
than twice the angular defect of triangle I. Indeed, the
first defect is

$$2R - (\alpha + \beta' + \gamma'') =$$
$$2R - \{\alpha + \beta' + \gamma'' + ((\alpha + \beta'' + \gamma') - 2R) + ((\alpha' + \beta + \gamma) - 2R)$$
$$+ ((\alpha'' + \beta + \gamma) - 2R)\},$$

on adding and subtracting the angles at A', B, and C. But
the sum on the right can be rearranged, on putting the
angles in each smaller triangle together, as

$$2R - (\alpha + \beta' + \gamma'') =$$
$$2R - \{\alpha + \beta + \gamma + (\alpha + \beta + \gamma) - 2R)) + ((\alpha' + \beta' + \gamma') - 2R))$$
$$+ ((\alpha'' + \beta'' + \gamma'') - 2R))\} =$$
$$2\{2R - \{\alpha + \beta + \gamma)\} - \{((\alpha' + \beta' + \gamma') - 2R)$$
$$+ ((\alpha'' + \beta'' + \gamma'') - 2R)\}$$

Since the last two terms
in brackets are negative,
by assumption, it follows that
$2R - (\alpha + \beta' + \gamma'') >$
$2\{2R - (\alpha + \beta + \gamma)\}$
as claimed.

amount by which its angle sum is less than two right angles) more than twice the angular defect of the original triangle. Repeating this construction more than doubles the defect each time, so a triangle can be constructed with a defect greater than two right angles, which is absurd. It follows that the original hypothesis of a triangle with an angle sum less than two right angles is absurd, and so the parallel postulate is proved.

The reader can be forgiven for feeling doubtful, although perhaps unsure where the mistake is (it is unmasked on page 122). Eventually Legendre came to feel dissatisfied with the "proof," and withdrew it in favor of yet another, which makes a different mistake (one we shall not be interested in). It is all these assaults on the "obvious" and their endless failures that is our point of concern. And there were indeed many more. The twenty-eight attempts mentioned above had been swollen by those of other authors than Legendre. In 1806 his French contemporary Joseph Louis Lagrange, the greatest mathematician in the world at the time, had the embarrassment of reading a paper on the parallel postulate to the Institut de France which so palpably assumed what it wanted to prove that the only response was a painful silence while Lagrange put his paper back in his pocket and the meeting moved on to next business.[3] And Pierre-Simon Laplace, the leading applied mathematician of his generation, took his stand on the observation that Newton's law of gravitation is so well attested by experience it must be regarded as true. It permits arbitrary scalings, and so the existence of similar, non-congruent figures must be allowed, and thus the parallel postulate is true. This is at best an argument that the world we live in can be described by Euclidean geometry; it does not show that the parallel postulate is necessarily true.

Perhaps more important, the mathematician Johann Lambert, a friend of Kant's, with whom he shares the honor of being one of the first to elucidate the nature of the Milky Way, had also written extensively on the parallel postulate. He took much of his information on the previous history of the topic from Klügel's thesis, and added some ingenious arguments of his own. But perhaps the most important thing was that he did not reach a definitive opinion – and that may in turn be the reason he did not publish his work. It was only published in 1786, nine years after his death.

Lambert agreed that the parallel postulate was equivalent to the existence of similar, non-congruent figures, and so in any geometry in which the parallel postulate is false there will be an absolute measure of length. He speculated that this was unacceptable on philosophical grounds, but decided these were too flimsy. He noted, as Legendre had done, that the argument that the angle sum of triangles cannot exceed two right angles is compelling, but the argument that it cannot be less is much weaker. He stated, without proof, that if it is assumed that the sum of the angles of every triangle is less than two right angles, then the angular defect $2R - (\alpha + \beta + \gamma)$ of a triangle is proportional to its area. This provoked him to a startling observation. He noted that on a sphere of radius r a triangle with angles α, β, and γ has area $r^2 (\alpha + \beta + \gamma - 2R)$, and said: "From this I should almost conclude that the hypothesis [that the sum of the angles of every triangle is less than two right angles] would occur in the case of an imaginary sphere." (Lambert [1786], § 82.) His argument is simply that on replacing r by $\sqrt{-1} \cdot r$, the formula becomes

$$-r^2 (\alpha + \beta + \gamma - 2R) = r^2 (2R - (\alpha + \beta + \gamma)).$$

Unfortunately, he gives no clue as to what a sphere of imaginary radius might really be, and the implications for physical space are nowhere discussed. The work ends with suggestions as to how the parallel postulate might be proved, but Lambert was clear that it had not been proved.

His correspondence with Kant shows him to have been more down-to-earth than the philosopher.[4] In his view, time and space were real, and not mere appearances. But, he wrote to Kant in 1770, "If you can instruct me otherwise, I shall not expect to lose much. Time and space will be *real* appearances. . . . I must say, though, that an appearance that absolutely never deceives us could well be something more than mere appearance." But he never shared with Kant the ideas that led him to realize that the parallel postulate was not secure. Since Kant had no doubt on that point, Lambert's reticence denied posterity an interesting exchange of views.

BOLYAI'S APPENDIX

Carl Friedrich Gauss

Enter the acutest critic of mathematics of his day, and one of its most profound innovators: Carl Friedrich Gauss. Gauss was a student at Göttingen under Abraham Gotthelf Kästner, and one of his student friends from 1796 to 1798 was Farkas Bolyai, father of János and himself a mathematician interested in the parallel postulate. Gauss was also a good, if ruthless correspondent, and much of what we know about this period on this subject we know from his letters, which have been kept and often published. For the same reason, it is difficult to know what Gauss understood about the parallel postulate, because he never drew his ideas together, and we are dependent on scattered remarks.

When Farkas Bolyai returned to his native Hungary in 1798 he kept in touch with Gauss. Gauss wrote to him hoping that they would stay in contact and Bolyai would tell him of his researches into the parallel postulate, but, Gauss went on: "Only, the path which I have chosen does not lead to the goal that one seeks, and which you assure me you have achieved, but rather makes the truths of geometry doubtful." [5] Undaunted, in 1804 Farkas Bolyai sent him the fruit of three years' thought about the parallel postulate: Gauss found a crucial error.

In 1816 Gauss criticized several of Legendre's arguments in a letter to Christian Ludwig Gerling. [6] In 1817, he wrote to the astronomer Wilhelm Olbers (after whom the paradox is named) to say: "I am becoming more and more convinced that the necessity of our geometry cannot be proved . . . . Perhaps only in another life will we attain another insight into the nature of space, which is unattainable to us now. Until then we must not place geometry with arithmetic,

Farkas Bolyai

which is purely *a priori*, but rather in the same rank as mechanics."[7] By 1829 he could write to Friedrich Wilhelm Bessel that "my conviction that we cannot base geometry completely *a priori* has, if anything, become even stronger," but that he probably would not publish his ideas in his lifetime because he feared the howls of the Boetians (a tribe the ancient Greeks had regarded as particularly stupid).[8] Bessel replied that he regretted this modesty on Gauss's part, and in April 1830 Gauss wrote back to say that "It is my inner conviction that the study of space occupies a quite different place in our *a priori* knowledge than the study of quantity. . . . we must humbly admit that if Number is the pure product of our mind, Space has a reality outside of our minds and we cannot completely prescribe its laws *a priori*." [9]

Gauss was insisting that there was no absolute logical necessity to geometry, that it could not be known *a priori* but had an arbitrary element in it, and could at best be a system of empirically valid truths akin to mechanics. But he did not want to publish his ideas, and the reason he gave need not be taken at face value. What indeed had Gauss learned about the parallel postulate in the years between 1804 and 1830?

Much of what he learned he had learned from others. The letter to Olbers, for example, was about the work of his former student Wachter, who had sent Gauss a defense of the parallel postulate, based on the idea that any four points in space, that do not all lie in a common plane, lie on a sphere. Evidently they met to discuss the problem, for Wachter then wrote to Gauss setting down what he recalled of their conversation. [10] In this letter he mentions the profound idea that if the parallel postulate is false then it will be necessary to study geometry on a sphere of infinite radius, but he did not spell out what such a mysterious object could be (we shall find out below). It is not clear if Wachter or Gauss came up with the idea orig-

Box 4. Schweikart's Memorandum

There are two kinds of geometry – a geometry in the strict sense – the Euclidean; and an astral geometry. Triangles in the latter have the property that the sum of their three angles is not equal to two right angles. This being assumed, we can prove rigorously:

a) That the sum of the three angles of a triangle is less than two right angles;
b) that the sum becomes ever less, the greater the area of the triangle;
c) that the altitude of an isosceles right-angled triangle continually grows, as the sides increase, but it can never become greater than a certain length, which I call the Constant. Squares have, therefore, the following form:

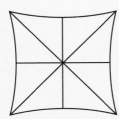

If this Constant were for us the radius of the Earth, (so that every line drawn in the universe from one fixed star to another, distant 90° from the first, would be a tangent to the surface of the earth), it would be infinitely great in comparison with the spaces which occur in daily life.

The Euclidean geometry holds only on the assumption that the Constant is infinite. Only in this case is it true that the three angles of every triangle are equal to two right angles: and this can easily be proved, as soon as we admit that the Constant is infinite.

inally. On the other hand, Gauss was quite delighted the next year, 1818, when Gerling passed on to him a note (see Box 4) written by a Professor of Jurisprudence at Marburg.[11] In it the author, F. K. Schweikart, set out the basic facts of what he regarded as a new geometry which could indeed be the true geometry of space.

Gauss replied, via Gerling, that he was uncommonly pleased with the note, which seemed to him to be almost entirely on the mark.[12] He disagreed about the astronomical implications of the mysterious constant, and went on to say that he could solve all the problems of astral geometry once the constant was known. As an illustration he gave a formula for the maximal area of a triangle in astral geometry in terms of the constant.

Matters were to be much more difficult for Gauss when Schweikart's nephew F. A. Taurinus got in touch in 1824. Gauss set out his views, but asked Taurinus to keep them confidential.[13] In 1825 Taurinus published a book explaining, to his satisfaction, why astral geometry, although logically consistent, was not a possible geometry of space. He particularly objected to the existence of an absolute measure of length.

János Bolyai

JÁNOS BOLYAI was born in Klausenburg, Transylvania, Hungary (now Cluj in Romania) on December 15, 1802, and moved with his parents to Maros-Vásérhely (now Târgu-Mures, Romania) in April 1804, when his father Farkas became professor of Mathematics at the Evangelical Reformed College. The town was a pleasant one in the wine district and the College dated back to 1557. Farkas adorned his house with pictures of Gauss, Shakespeare, and Schiller, and became something of a playwright himself; he also translated several English and German works into Hungarian.

He brought his son János up in the manner suggested by Rousseau's *Emile*, which emphasized the importance of play and naturalness against the prevailing authoritarianism of the age. János read the first six Books of Euclid's *Elements* under his father's direction, and in due course moved on to Euler's *Algebra*, before persuading his father to let him attend lectures at the College; he was then only twelve years old. By the time he graduated at the top of the class, in 1817, he had won prizes for Latin and was acclaimed as a violinist. He had even taught the other students in mathematics and physics. But unlike his father he had no taste for poetry. His father thought he was, if anything, inclined to study too much, at some cost to his health. From 1818 to 1823 János studied at the Royal Engineering Academy in Vienna, which trained

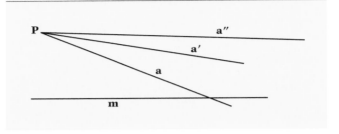

Figure 3. The line parallel to *m* through *P* is the limiting position of lines through *P* that meet *m*

cadets for military service. He then served as an engineer in the Austrian Army for ten years, and this seems to have given him some time for mathematics. In 1833, weary of the military life, he retired on a pension as a semi-invalid to Maros-Vásérhely.

While in the army, his interests had turned to the parallel postulate, doubtless because of his father. In Vienna he had fallen in with Carl Szász, who suggested to him that the line through a point *P* parallel to a line *m* might be considered as the limiting position, *a''*, of a line *a* through *P* as it rotates (see Figure 3). It has the property of being the first line such that every line below it meets the line *m*. Bolyai later called it an asymptotic parallel.

By 1820 he began to think that his failures to prove the parallel postulate might arise because in fact the parallel postulate was not true. He switched direction and henceforth attempted to show that there could be a geometry independent of the parallel postulate. He wrote to his father, in words that recall his Rousseauist upbringing, that, "one must do no violence to nature, nor model it in conformity to any blindly formed chimera; that on the other hand, one must regard nature reasonably and naturally, as one would the truth, and be contented only with a representation of it which errs to the smallest possible extent."

His father was alarmed, and wrote back:

> You must not attempt this approach to parallels. I know this way to the very end. I have traversed this bottomless night, which extinguished all light and joy of my life. I entreat you, leave the science of parallels alone . . . I thought I would sacrifice myself for the sake of the truth. I was ready to become a martyr who would remove the flaw from geometry and return it purified to mankind. I accomplished monstrous, enormous labours; my creations are far better than those of others and yet I have not achieved complete satisfaction. . . . I turned back when I saw that no man can reach the bottom of this night. I turned back unconsoled, pitying myself and all mankind. Learn from my example: I wanted to know about parallels, I remain ignorant, this has taken all the flowers of my life and all my time from me.[14]

And yet again:

> I admit that I expect nothing from the deviation of your lines. It seems to me that I have been in

these regions; that I have travelled past all reefs of this infernal Dead Sea and have always come back with broken mast and torn sail. The ruin of my disposition and my fall date to this time. I thoughtlessly risked my life and happiness – *aut Caesar aut nihil* [either Caesar or nothing].[15]

Happily, the son did not listen to his father, and on November 3, 1823 he could write to say that he was succeeding:

I am determined to publish a work on parallels as soon as I can put it in order, complete it, and the opportunity arises. I have not yet made the discovery but the path that I am following is almost certain to lead to my goal, provided this goal is possible. I do not yet have it but I have found things so magnificent that I was astounded. It would be an eternal pity if these things were lost as you, my dear father, are bound to admit when you see them. All I can say now is that I have created a new and different world out of nothing. All that I have sent you thus far is like a house of cards compared with a tower. I am as convinced now that it will bring me no less honor, as if I had already discovered it.[16]

His father advised him to publish his results as soon as possible, as an appendix to a work on geometry that he had been writing for some time. János later commented that:

BOLYAI'S APPENDIX

He advised me that, if I was really successful, then there were two reasons why I should speedily make a public announcement. Firstly because the ideas might easily pass to someone else who would then publish them. Secondly there is some truth in this, that certain things ripen at the same time and then appear in different places in the manner of violets coming to light in early spring. And since all scientific striving is only a great war and one does not know when it will be replaced by peace one must win, if possible; for here pre-eminence comes to him who is first.[17]

When János visited his father in February 1825, he was unable to convince him, worried as he was about an arbitrary constant that entered the formulae his son had found. As late as 1829 they continued to disagree about what the younger man had done, and his father continued to advise János not to waste his life like a hundred geometers before him had. Finally, they agreed to publish it anyway. The two-volume work, entitled *Tentamen juventutem studiosam in Elementa Matheosis purae,* [. . .] was published by the College in Maros-Vásérhely in 1832. A copy was sent to Gauss, but it was lost in the chaos of a local cholera epidemic, and another was sent.

On March 6, 1832, Gauss replied:

If I commenced by saying that I am unable to praise this work, you would certainly be surprised for a moment. But I cannot say otherwise. To praise it, would be to praise myself. Indeed the whole contents of the work, the path taken by your son, the results to which he is led, coincide almost entirely with my meditations, which have occupied my mind partly for the last thirty or thirty- five years. So I remained quite stupefied. So far as my own work is concerned, of which up till now I have put little on paper, my intention was not to let it be published during my lifetime. Indeed the majority of people have not clear ideas upon the questions of which we are speaking, and I have found very few people who could regard with any special interest what I communicated to them on this subject. To be able to take such an interest it is first of all necessary to have devoted careful thought to the real nature of what is wanted and upon this matter almost all are most uncertain. On the other hand it was my idea to write down all this later so that at least it should not perish with me. It is therefore a pleasant surprise for me that I am spared this trouble, and I am very glad that it is just the son of my old friend, who takes the precedence of me in such a remarkable manner.[18]

Farkas was pleased that the great geometer had endorsed his son's discoveries, but the son was appalled. It was to be almost a decade before he could be convinced that his father had not confided the ideas in the *Appendix* to Gauss, who then dishonestly claimed priority, just as his father had warned him that someone might. A long period ensued when father and son did not speak to each other. Farkas disapproved of the fact that János lived unmarried with a woman (Rosalie von Orban) by whom he had three children. Father and son then resumed an uneasy relationship which lasted until Farkas died in 1856, and János's relationship with Rosalie ended about the same time. János died in 1860.

János Bolyai's New Geometry

WHAT DID JÁNOS's new world from nothing consist of? He began by defining parallels in the way Carl Szász had suggested. (See Figure 4.)

If lines *AM* and *BN* lie in the same plane and *AM* is not cut by *BN*, but every line in the angle *ABN* cuts *AM* then Bolyai said the line *BN* is *parallel* to the line

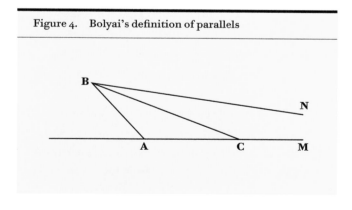

Figure 4.   Bolyai's definition of parallels

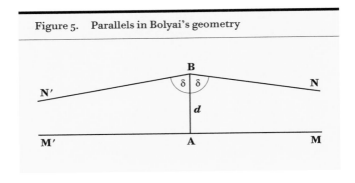

Figure 5.   Parallels in Bolyai's geometry

AM. There is only one line through the point *B* parallel to *AM* to the right, and there is another in the other direction (shown as *BN'* in Figure 5).

In this figure we have made the angles at *A* right angles. It follows that the angles *NBA* and *N'BA* are equal to an amount δ say, which Bolyai assumes is less that a right angle. Later Bolyai will exploit this definition to investigate how the angle δ depends on the length of the perpendicular *BA*.

Bolyai then established some basic properties of parallel lines based on his new definition. In particular he showed that if *a* and *b* are parallel and *A* is a fixed point on *a*, then there is a unique point *B* on *b* such that the angles *MAB* = α and *NBA* = β are equal (see Figure 6).

Bolyai now made an important shift into three dimensions. This is a novel feature of his geometry, quite unlike previous investigations of the parallel postulate, and quite unlike Euclid's *Elements* for that matter. It can be appreciated by recalling that when

Figure 6.   Parallels and equal angles in Bolyai's geometry

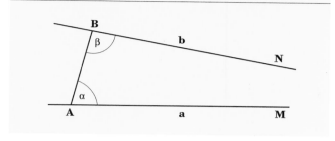

we wish to study the geometry on a sphere, we naturally think of the surface of a sphere sitting in space.
We speak of distances between pairs of points on this
sphere, by which we mean the distance between the
points measured along the surface of the sphere (no
tunnelling allowed). We call this geometry on the
sphere the "induced" geometry on the sphere, by
which we mean that it has inherited its concept of distance from the surrounding Euclidean space.

In the same way, but starting with non-Euclidean
three-dimensional space, Bolyai introduced a special
surface, which he denoted by the letter F, obtained
as follows. He took a straight line a with a point on it,
A, and in any plane containing the line a he considered all the parallel lines to a (in the same direction,
which is to the right in Figure 7). On each parallel
line, b, he located the point B such that the angles
MAB and NBA are equal (their size depends on the
position of the point B). This gave him a curve, which
he denoted L, in the plane containing the lines a and

Figure 7. Bolyai's *L*-curve

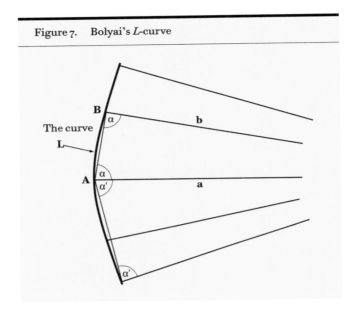

The curve L

b, and then, as the plane through *a* is varied, the surface *F*. The names may have been derived from the German *Linie* for line and *Fläche* for surface.

The surface *F* is bowl-shaped, and has the original line *a* as an axis. The surface *F* can be thought of as swept out by the curve *L* as it is rotated about the axis *a*. (See Figure 8.)

Bolyai observed that if the parallel postulate is true, then the angle δ in Figure 5 is a right angle, the curve *L* is just the straight line through *A* perpendicular to *a* and the surface *F* is just the plane through *A* perpendicular to *a*. But if the parallel postulate is false, then the angle δ in Figure 5 is less than a right angle, the curve *L* is not a straight line but a curve

Figure 8.   Bolyai's *F*-surface

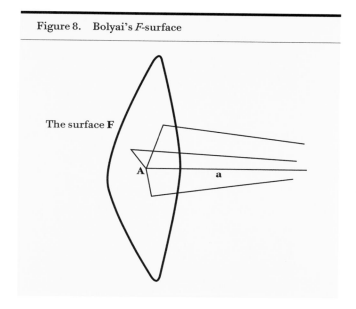

The surface **F**

A

a

perpendicular to *a* and the surface *F* is not the plane but a surface perpendicular to *a*. He now set out to discover as many theorems as possible whichever option is true, noting along the way theorems which were only true if the parallel postulate is false.

Among the theorems he discovered that hold when the parallel postulate is false are these:

1. All the lines *b* parallel to the axis *a* meet the surface *F* at right angles (so the surface F could plausibly be called a sphere of infinite radius such as Gauss and Wachter might have had in mind); and

2. Any plane containing the axis *a* meets the surface *F* in a curve *L*.

These theorems are also true when the parallel

postulate is true, but they are trivial because the surface F is a plane in that case. János also showed:

3. Any plane not containing the axis a meets the surface F in a circle. (This theorem is only true when the parallel postulate is false.)

The most important result János established is true whether the parallel postulate is true or false, but it is much more interesting, even surprising, when the parallel postulate is false:

4. On any surface F, if two curves L cross a third and the sum of the interior angles is less than two right angles, then the two L curves intersect.

This last property is the equivalent of the parallel postulate for curves L on a surface F. It follows that if the L curves are taken as straight lines, then Euclidean geometry holds on the surface F, whether or not the parallel postulate is assumed. Bolyai called this result an "absolute" theorem, one that is true in a geometry in which the parallel postulate is true and also in a geometry in which the parallel postulate is false. It means that whenever Bolyai is studying a problem in his new geometry, he can relate it in a natural way to a problem in Euclidean geometry.

It also means that in a space in which the parallel postulate is false there is a surface on which the induced geometry is Euclidean. Mathematicians had failed to find a geometry in Euclidean three-dimensional space upon which the induced geometry was non-Euclidean, so this discovery of Bolyai's was striking. It allowed him to compare figures in two-dimensional non-Euclidean geometry with other figures in two-dimensional Euclidean geometry, much as we can pass between figures on a sphere and corresponding figures on a plane.

This remarkable theorem means that the geometer can study geometry on the sphere by thinking of it in a three-dimensional Euclidean space, and two-dimensional Euclidean geometry by thinking of it in a three-dimensional non-Euclidean space. It leaves open the question of finding a surface in three-dimensional Euclidean space on which the induced geometry would be non-Euclidean geometry. Had such a surface been easy to find, the discovery of non-Euclidean geometry would doubtless have been made early on. In fact, it was to turn out that no such surface exists. That does not mean that non-Euclidean geometry does not exist, only that there is no intuitively accessible model of it as a surface in Euclidean space. János Bolyai's discovery means that a three-dimensional non-Euclidean geometer would find Euclidean geometry as easy to study as we find spherical geometry.

To compare triangles in non-Euclidean geometry with those on the F-surface, Bolyai considered a (non-Euclidean) triangle ABC, right-angled at B, and the lines AM perpendicular to ABC, BN, and CP parallel to AM. (See Figure 9.)

Bolyai now moved further away from Euclid's way of thinking about geometry: he introduced trigonometric formulae. He considered the surface consist-

Figure 9.   Euclidean and non-Euclidean triangles

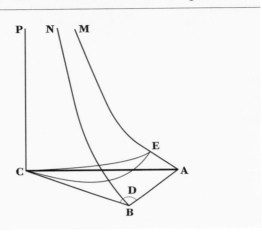

ing of all the lines parallel to BN and CP through
points of BC, and showed that it is perpendicular to
the similar surface consisting of all the lines parallel
to BN and AM through points of AB. He then took
the F-surface through C perpendicular to CP, and
supposed that BN meets it at D and AM meets it at E.
He then deduced that the angle between the L-curves
CD and DE at D is a right angle. It is also clear that
the angle between the L-curves DC and EC at C is
equal to the angle between the non-Euclidean lines
BC and AC at C. Moreover, rotating the figure about
the axis CP shows that the ratio of the circumferences
of circles whose radii are CD and CE in the F-surface
is equal to the ratio of the circumferences of the circles whose radii are CB and CA in the non-Euclidean
plane. This means that the elementary trigonomet-

ric formulae connecting the sides and angles of the *F*-triangle *CDE* can be transformed into results about the sides and angles of the corresponding non-Euclidean triangle. But note that at this stage Bolyai did not have a formula for the circumferences of a non-Euclidean circle of a given radius.

Bolyai next considered the formulae that relate the sides and angles of a triangle on the (Euclidean) sphere, which in his day were taught to every land surveyor and military engineer. In particular, he showed that they too are absolute theorems – they are true whether the parallel postulate is true or false, which is a rather remarkable result. It meant that he could also use spherical trigonometry in the new setting.

To find the appropriate trigonometric formulae in a geometry where the parallel postulate is false Bolyai began with the fundamental figure in such a geometry. This consists of a straight line *AB* of length *y*, meeting a line *AM* at right angles and a line *BN* at an acute angle *u*, where the lines *AM* and *BN* are par-

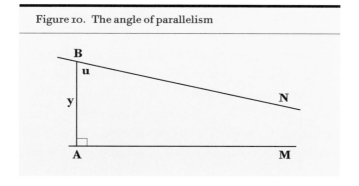

Figure 10. The angle of parallelism

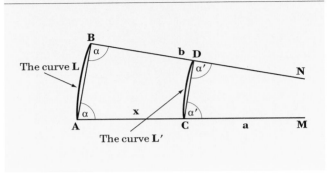

Figure 11. The curve L'

allel. The length y determines the angle u and vice versa. The angle u is called the *angle of parallelism* corresponding to the length y, and is sometimes written $u = \Pi(y)$. Bolyai now sought (in § 29) the expression for the length of the line segment y as a function of the angle u (Figure 10).

His argument was a long one. He first considered an arbitrary point C on the line a and the L-curve through it. He called this L-curve L'. (See Figure 11.)

He showed that the ratio $AB{:}CD$ is independent of AB and depends only on the length $AC = x$. He denoted this ratio X and set himself the task of evaluating it. He gave his answer in the form of a formula relating u and Y, where Y is the same function of y that X is of x, and u is the angle of parallelism corresponding to the length y. He showed by a simple scaling argument that given lengths x and x' the corresponding X and X' satisfy $Y^{1/y} = X^{1/x}$, from which it

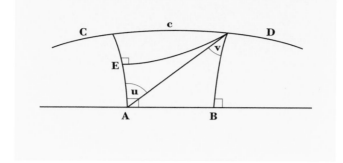

Figure 12. The angles u and v

follows, although Bolyai did not say so immediately, that $X = e^{kx}$ for some arbitrary constant k.

To find X he considered the curve which is everywhere equidistant from a straight line. In Euclidean geometry this is another straight line, but if the parallel postulate is false it will not be. Bolyai imagined it was swept out by the tip of a line segment that moves perpendicular to the given line. In Figure 12 the triangle ABC is supposed to slide rigidly along the line a, with the edge AC remaining always perpendicular to a. The point C draws the equidistant curve, c, to the line a. The segment BD is another position of the segment AC, so the point D lies on the curve c. The angles u and v are as shown: $u = CAD$ and $v = ADB$.

Bolyai showed that the ratio of the length of the line segment AB to the length of the segment CD of the curve c is equal to $\sin(v) / \sin(u)$ by dropping the perpendicular DE from D to AC and first applying

Figure 13. The ratio $X = AB/CD$ where $x = AC$

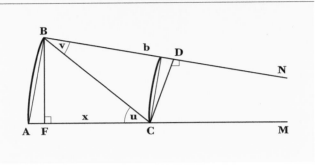

his trigonometric formulae to the triangles AED and ABD. Then he used a limiting argument to find the ratio of the lengths AB and CD. He observed that since the ratio AB/CD is clearly a constant (it depends only on the height AC not the width AB) this ratio can be evaluated in the limit as AC moves off infinitely far, when the angle u tends to a right angle and the angle v to the angle δ (compare Figure 5 above).

However, he also showed (see Figure 13) that the ratio X he was interested in before was equal to $\sin(u)/\sin(v)$. From this he could deduce the result he wanted: the angle of parallelism is given by the formula $Y = \cot(u/2)$. If we use the result that $Y = e^{ky}$ this can be re-written as the formula $\sinh y = \cot u$ or as $\sinh y = \cot \Pi(y)$.

From this formula Bolyai proceeded rapidly to deduce a number of useful results. He found the formula for the perimeter of a circle of radius r, and then the formulae connecting the sides and angles of triangles when the parallel postulate is false. It is these formulae that contain an arbitrary constant and that had accordingly worried his father. Bolyai observed that when the parallel postulate is false any three pieces of information about a triangle suffice to determine the rest, and he gave all the formulae in the easy case when the triangle has a right angle (formulae for the general triangle follow with a little routine work). This is not quite the case in Euclidean geometry: knowing the angles of a triangle determines its shape, but not its size (one might of course reply that only two pieces of information are required to determine the angles of a Euclidean triangle, because the angle sum must be two right angles). Bolyai also showed that the new formulae reduce to their Euclidean equivalents when the arbitrary constant increases to infinity. In particular, the formula corresponding to Pythagoras's Theorem in non-Euclidean geometry reduces to the usual expression for Pythagoras's Theorem in Euclidean geometry in this way, showing that small regions of non-Euclidean space are approximately Euclidean.

Bolyai went on to show that in the new geometry the area of a triangle in the plane (not on the special surface F) is equal to its angular defect, and then to show that if the parallel postulate is false one can construct a square equal in area to a given circle. Thus Bolyai squared the circle. Squaring the circle is synonymous with achieving the impossible and had been regarded that way since at least 450 B.C.E. when the

Greek playwright Aristophanes wrote *The Birds*. So it must have been with some excitement that Bolyai realized that in his new geometry one can indeed square the circle, and he addressed this result (in § 33) to "the friends of truth" who, he supposed, would not find it unwelcome. His method occupies the final sections of the book (§§ 34-43), and can be summarised in the following way.

How Bolyai Squared the Circle

THE CONSTRUCTION proceeds in two stages. First, a particular circle is drawn whose area is πk^2. Second, a square with the same area is constructed; it will be one with angles of $\pi/4$ at each vertex. Because the second of these constructions is routine, Bolyai would have focussed his attention on the first of these stages, which is ingenious. To see why it works, observe that Bolyai had shown that the area of a circle of non-Euclidean radius r is $4\pi k^2 \sinh^2(r/2k)$. So Bolyai needed to construct a length, r, such that $\sinh^2(r/2k) = 1/4$. This can be done in non-Euclidean geometry by constructing a suitable angle, because lengths and angles are related by the formula for the angle of parallelism, which says $\sinh y = 1/\tan\prod(y)$. So the problem is reduced to the construction of a length y such that $\tan\prod(y) = 2$. Bolyai solved this construction by a variant on the construction for parallels: if at opposite ends of a segment of length r parallels are drawn, meeting the segment at equal angles z, then $\tan z = 2/\tan\prod(r/2k)$ (this construction is described in "A" below). Bolyai chose $z = \pi/4$, so that $\tan z = 1$, thus constructing a length r for which $\tan\prod(r/2k) = 2$. With this value of r, the area of the circle is πk^2, as required.

The clever part is to show that the lengths and angles can all be constructed by elementary means, using just those construction methods which are valid in both Euclidean geometry and non-Euclidean geometry.

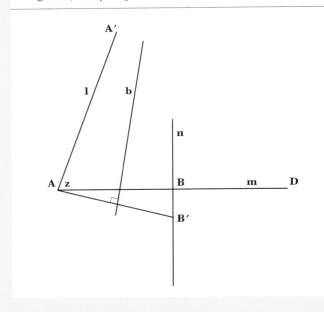

Figure 14. Bolyai squares the circle: Construction A

A. Given an angle z to construct a segment of length r such that $\tan z = 2/\tan\prod(r/2k)$.

   1) Given lines l and m meeting at the point A at the angle $DAA' = z$.

   2) Draw a line n perpendicular to $m = AD$ and parallel to AA' (this construction is described in "B" below). Let n meet m at the point B.

   3) Draw the line b bisecting the strip ln (this construction is described in "C" below).

   4) Draw the perpendicular from A to b and let it meet n at B'. The segment AB' has length r.

Figure 15. Bolyai squares the circle: Construction B

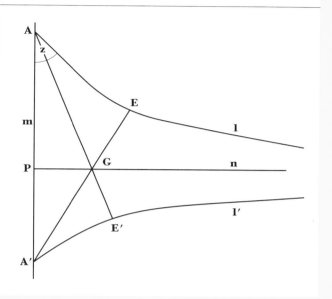

B. Given lines *l* and *m* meeting at the point *A* at an angle *z*, we construct a segment *AP* on *m* and a line *n* through *P* at right angles to *m*, such that the line *n* is parallel to the line *l*, i.e. such that *z* is the angle of parallelism corresponding to *AP*.

1) Pick a point *A'* on *m* such that the parallel *l'* to *l* through *A'* makes an acute angle with *m* (as shown).

2) Drop the perpendicular *A'E* from *A'* to *l*.

3) Drop the perpendicular *AE'* from *A* to *l'*, and let it meet the line *A'E* at *G*.

4) Drop the perpendicular *GP* from *G* to *m*.

The line *GP* is the required line *n,* and the segment *PA* is the required segment at right angles to *n,* because the perpendicular *GA* to *PC* is also parallel to *l.* The construction works because the perpendiculars from the vertices of a non-Euclidean triangle to the opposite sides meet in a point, even when one of the vertices is "at infinity."

C. To draw a line bisecting a strip between two parallel lines *l* and *n.*

 1) Pick a point *A* on *l* and a point *B* on *n* and join them by a line, *k.*

 2) Bisect the angles between *k* and *l* at *A* and between *k* and *n* at *B* and let them meet at *P.*

Figure 16. Bolyai squares the circle: Construction C

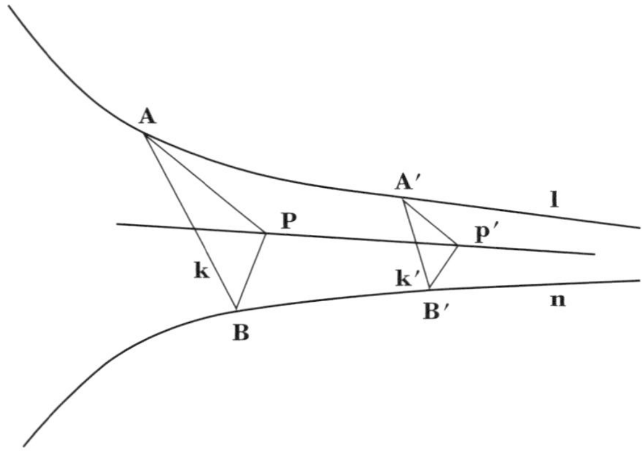

3) Repeat this construction with new points A' on l and a point B' on n to obtain P'.

The line PP' is the bisector of the strip between l and n.

The construction works because the bisectors of the angles of a non-Euclidean triangle meet in a point, even when one of the vertices is "at infinity."

For completeness we add Construction D.

D. Given a line l and a point P not on it, we construct the parallel to l through P.

1) Drop the perpendicular PQ from P to l, and let it meet l at Q.

Figure 17. Bolyai squares the circle: Construction D

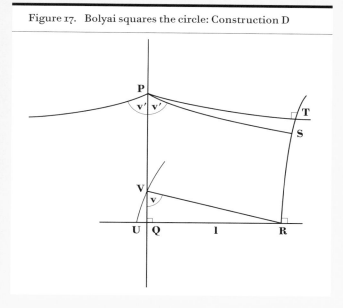

2) Pick a point R on l and draw the perpendicular RS to l on the same side of l as PQ

3) Drop the perpendicular PT from P to RS. Note that in the quadrilateral $PQRT$ the angles at Q, R, and T are right angles and the angle at P is acute.

4) Draw a line segment RU on l the same length as TP. Note that U will lie the other side of Q from R.

5) Draw the circle center R and radius RU and let it meet QP at V. Denote the angle RVQ by v.

The line through P making an angle v with PQ (and on the opposite side to T) is the parallel to l through P. The best proof of this is analytic, and we omit it.

Now we indicate how Bolyai constructed a square with four angles of $\pi/4$. He observed that it is enough to construct a triangle ABC with angles of $\pi/2$, $\pi/4$ and $\pi/8$, and then put eight copies of it together. Thereafter, his own description was cryptic in the extreme. First a length X is produced "by mere square roots" and from it another, x, is determined. The second step, a reference to the formula for the angle of parallelism and the way in which it can be used in constructions, shows that $\cosh(x/k) = 1/\sin\prod(x/k)$. On the other hand, from looking at the right-angled triangle formed from two copies of triangle ABC and applying Pythagoras's Theorem we can deduce that if the square has sides of length s and diagonals of length $2d$ then $\cosh(s/k) = \cosh^2(d/k)$. A cosine formula applied to the angle $\alpha = \pi/8$ allows us to write $\cos\alpha = \tanh(s/2k)/\tanh(d/k)$, and a short calculation then shows that $\tanh(s/k) = \sqrt{\sqrt{2}/2}$. Arguments of this sort are presumably what Bolyai had in mind. Bonola indicated another construction, which proceeded by applying another trigonometric formula to triangle ABC to reduce the construction to the construction of segments b' and c' respectively corresponding (via the angle of parallelism formula) to the angles $\cos\alpha = \sin3\alpha$ and $\sin2\alpha = \sin(\pi/4) = \sqrt{2}/2$.

The required length is then the third side of a right-angled triangle with side b' and hypotenuse c'.

The constructions raise the question of what other squares can be constructed. Bolyai observed that it can be done whenever $\tan^2 z$ is an integer, and whenever it is a rational number whose denominator is product of distinct primes of the form $2^n + 1$. This was because he could adapt Gauss's construction of the regular Euclidean n-gon, where n is of the form $2^n + 1$ to the new setting. It is perhaps ironic to note that he described Gauss's work as a "remarkable invention of our, nay of every age." He was to change his opinion of the man, at least, quite soon.

Anticlimax

THE IMMEDIATE RECEPTION of Bolyai's work was poor indeed. It was indeed unlikely that an Appendix to a two-volume work in Latin written by a little-known Hungarian would command much attention. Gauss did not choose to draw attention to it, and so the international community passed it by, never to notice it in the lifetimes of either the father or the son.

The same was true of the remarkably similar achievement of the Russian mathematician Nikolai Ivanovich Lobachevskii.[19] He had followed a similar career to János Bolyai's, becoming a professor at the University of Kasan. In 1829 he published the first of several articles in the *Kasan Messenger* describing an alternative geometry to Euclid's. Lobachevskii deeply believed that the Euclidean foundations were flawed. In his view terms like "line," "surface," and "position" were obscure, and certainly not fundamental. Rather, geometry should be based on ideas about bodies and the motion of bodies. This echoes the ideas of d'Alembert in the *Encyclopédie*, whose philosophy of geometry was well adapted to the application of geometry and the calculus to Newtonian mechanics. Lobachevskii argued that ideas about straight lines are derived from a consideration of bodies, and when this is done carefully, lines need not be as Euclid had said they were.

In particular, Lobachevskii showed that in the new geometry the angle sum of a triangle is always

Nikolai Ivanovich Lobachevskii

less than two right angles, and the angle sum gets less as the triangle gets bigger. This suggested to Lobachevskii that one could attempt to see if space was in fact non-Euclidean. His idea was to consider the par-

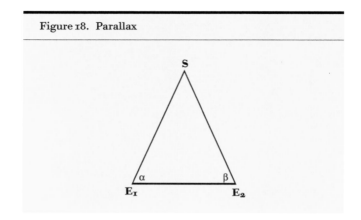

Figure 18. Parallax

allax of stars. (See Figure 18.) In Euclidean geometry the parallax (measured by $\pi - \alpha - \beta$ in the figure) gets smaller as the star is further away, becoming arbitrarily small if stars are to be found far enough away. But in the new geometry the parallax cannot fall below a certain level, determined by the diameter of the Earth's orbit, so if measurements of parallax always exceed a certain non-zero amount, this would be evidence in favor of the validity of the new geometry. Lobachevskii took advantage of the latest astronomical work to consider various stars, but found the observations were inconclusive.

Sadly for Lobachevskii, he had no skill as a writer, and his papers were unconvincing. They are a string of deductions from a hypothesis that differs from the parallel postulate, and as such beg the question of whether this can consistently be done. He published in French and German in an effort to be read internationally, but could never show that his new geometry was not self-contradictory and the new geometry not an illusion. Russian authorities like Mikhail Vasil'evich Ostrogradskii denigrated his work, which they did not understand. His little German book got only one review, which was a travesty. But Lobachevskii also sent a copy of it to Gauss in Göttingen. He immediately acclaimed Lobachevskii's work, read the earlier papers in Russian, and in 1842 had him made a corresponding member of the Göttingen Academy of Sciences. This was to be the only acclaim Lobachevskii was to receive in his lifetime.

What gave Lobachevskii his conviction that his geometry was not self-contradictory was his finishing point. Lobachevskii expressed his theorems in the language of trigonometry and the calculus. He deliberately sought out formulae because he deeply believed that geometry was about measurement, and that measurements, numbers, are related to one another by formulae. In turn, the validity of these formulae was a matter of algebra, whatever might be their geometrical significance.

In fact, his papers and books closely resemble János Bolyai's. Both worked in three dimensions. Both expressed their fundamental results in terms of new trigonometric formulae. Lobachevskii is clearer in some respects, and more concerned to find theorems in the new geometry than to find theorems common to both Euclidean and the new geometry. He had the sensible habit of giving names to important features

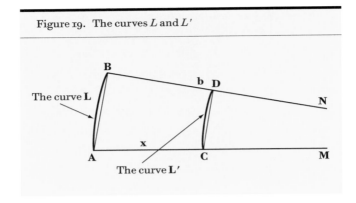

Figure 19. The curves L and L'

The curve **L**

The curve **L'**

of the new geometry (which he called the imaginary geometry) whereas Bolyai merely used letters (and was criticized on this account by Gauss). Lobachevskii called the curve perpendicular to a family of parallel lines a horocycle, and the corresponding surface a horosphere. Like Bolyai, he showed that the geometry on the horosphere is Euclidean.

The Bolyais eventually heard of Lobachevskii's work in 1844 from a Hungarian mathematician and physicist, Franz Mentovich, who met Gauss in Göttingen. Unable to obtain any copies of Lobachevskii's work, Farkas Bolyai wrote to Gauss in 1848, who replied that they might do better trying to locate the German booklet Lobachevskii had published in 1840. They got a copy (on October 17, 1848). They found numerous small points to criticize. Some of their arguments were preferable, notably the proof of the exponential relationship between lengths along two L-curves with the same axes. (See Figure 19.)

But they were rightly impressed by the Russian's derivation of the trigonometric formulae. At first the two Hungarians thought of publishing a reply, but their efforts petered out, and it must be a sad reflection on their state that by now they were willing to give up. They had discovered a new geometry, one of the most momentous discoveries ever made, and the world simply ignored them.

Acceptance

GAUSS DIED ON FEBRUARY 23, 1855. After his death mathematicians started to work through the many notebooks he had left, guided by the clues in his correspondence, and when appropriate that too was published. It turned out that he had known a great deal that he had not published. The best explanation for this unusual habit may lie in the right interpretation of his own aphorism *pauca sed matura* – "little, but ripe." It does not mean that he only published his best work. Indeed, he published a large amount of rather routine calculations in astronomy, which was congenial work for him and significant in the discovery of the larger asteroids. Rather, it means that he held off publication in pure mathematics until he felt he had the right organizing idea for a body of work. Without that idea he would continue to work until he was satisfied that he had thought through the matter fully. Sartorius von Waltershausen, who knew Gauss personally and published a memoir about him in 1856, said that "Gauss always sought to give his researches the form of a completed work of art . . . and would therefore never publish a work before it had attained this completely worked-out form that he desired. After a good building has been put up, he used to say, you can no longer see the scaffolding." [20]

In the case of non-Euclidean geometry, mathematicians found many notes about it among his papers, as well as the books by Bolyai and Lobachevskii, and letters confirmed that for many years Gauss had thought that a new geometry was possible. He refrained from publishing either because the right idea never came to him (not even through the post) or because he was truly afraid of the clamor such a discovery would set off, or perhaps because the priority had already gone to someone else. It was indeed the case that neither he, nor the Bolyais, nor Lobachevskii, had precisely shown that a contradiction would never be found. But psychological conviction was there in abundance, and with Gauss's endorsement confidence in the new geometry began to spread. Sadly, Lobachevskii died on February 24, 1856, Farkas Bolyai on November 20, 1856 and his son János on January 27, 1860, and none of them ever benefited from the new appreciation of their work.

A new climate of sympathy for non-Euclidean geometry, and the specific comments about it that Gauss left behind, helped mathematicians find the neglected work of its discoverers. Some of Lobachev-

skii's was already in French and German. His booklet of 1840 was the best account, and it was translated
into French by Guillaume Jules Hoüel in 1866. Hoüel
followed it with a translation of the *Appendix* by Bolyai
the next year, and in 1869 set a translation of an essay
by Eugenio Beltrami before a French audience. Meanwhile the *Appendix* had been translated into Italian
by Giuseppe Battaglini in 1868. Gradually their work
could spread, but not before its imperfections were
removed and the lingering threat that a hidden contradiction might surface was finally dispelled. For this
a radical re-think was required, which even Gauss had
arguably failed to produce.

The mathematician who first had the idea that
Gauss "missed" was a student of his, Bernhard Riemann. Riemann is undoubtedly the most important
mathematician of the mid-nineteenth century, although his entire works fill only one volume. In 1854
he presented his *Habilitationsschrift* at Göttingen
(the qualification that permitted one to teach at a German university, but not necessarily to be paid). He
had offered three titles, as the rules required. To his
surprise, although he should have known better,
Gauss chose the topic on the foundations of geometry. The thesis was defended within the Philosophy
Faculty, because mathematics was a part of that Faculty, which may account for the absence of formulae;
unfortunately this makes it harder to understand. But
in outline Riemann took one of Gauss's best results
and made it truly fundamental – exactly what Gauss

Bernhard Riemann

most liked to do. Gauss commented to a friend afterwards that he had seldom been so excited as by the profundity of Riemann's ideas.

Gauss's result had been that on a surface there is a way of defining its curvature at each point which is intrinsic: it can be determined from measurements taken entirely in the surface and which do not involve stepping off the surface and into space. He called this result his *Theorema Egregium* or exceptional theorem, and it is one of the key results in his book of 1827 on differential geometry. In Gauss's day, differential geometry was the study of curves and surfaces in space by means of the calculus. The idea that there was an intrinsic quantity was entirely novel, although the simplest illustration is mundane. The curvature at every point of the plane is zero (it is said to be flat) and so is the curvature of a cylinder. Small patches on a cylinder can be rolled out flat without any distortion: that is exactly what old-fashioned printing does.

Riemann's idea was to take the concept of curvature and to argue that geometry was fundamentally about two types of problem: the intrinsic properties of a surface, and the ways in which a surface can be mapped into another space. But Riemann did not stop with surfaces. He introduced the idea of n-dimensional spaces – rather vaguely, to be sure – as spaces where n coordinates were needed to specify the position of a point, and where it was possible to measure lengths along curves. He indicated how the Gaussian idea of intrinsic curvature could be generalized to this new setting. And he mentioned, almost in passing, that there were three two-dimensional geometries where the curvature was constant: the cylinder (curvature zero), the sphere (curvature positive), and surfaces of constant negative curvature (which he only alluded to).

Examples of spaces of constant negative curvature had been given in the literature, but no-one had connected them to non-Euclidean geometry. Indeed, Riemann only did so rather obscurely. He opened his account with a reference to a darkness that had overlain geometry from Euclid to Legendre. And he observed that in a geometry with constant curvature the angle sum of a triangle was known in all cases once it was known in one. This cryptic utterance means that either the angle sum is always two right angles, or it is always more, by an amount that varies from triangle to triangle, or it is always less, again by an amount that varies from triangle to triangle. This was all he felt he needed to say about the possibility of a non-Euclidean geometry from his point of view.

So Riemann broke entirely with the idea that geometry is about undefined but clear ideas such as point, line, and plane. For him, the basic terms are point and distance. There are infinitely many geometries in each dimension, although not many of the simplest kind, those with constant curvature. Among these geometries, Euclidean geometry is in no way fundamental. It is just one among many. There is no need to define a two-dimensional geometry by first exhibiting it as the geometry on a surface embedded in Euclidean three-dimensional space. At a stroke, the failure of Bolyai and Lobachevskii to present non-Euclidean geometry in that way became irrelevant. It only showed that two-dimensional non-Euclidean geometry might not be made visible in that way, not that such a thing was actually impossible.

Riemann's ideas circulated slowly. They were published for the first time only in 1868, after his death two years before. By then news of them had reached the Italian mathematician Eugenio Beltrami, and they gave him the courage to defy the lingering doubts of his supervisor and publish his own account of non-

Eugenio Beltrami

Euclidean geometry. With this, mathematicians for the first time had an impeccable proof of its existence, and the acceptance of non-Euclidean geometry was assured.

Beltrami began with the familiar map of a hemisphere onto a plane, which is obtained by imagining a point source of light at the center of the sphere. The map associates to each point P of the hemisphere its shadow P' on the plane. (See Figure 20.)

The map has the agreeable property of mapping great circles on the hemisphere (which are geodesics, curves of shortest length joining any two of their points) to straight lines, which are geodesics in the plane. It is called geodetic projection for that reason.

What of the distances? Plainly these are distorted. The distance between P' and Q' depends not only on the distance between P and Q on the sphere, but on where the points P and Q are: the shadow of points near the equator is stretched much more than the shadow of points the same distance apart but nearer a pole. There is a formula one can write down, which sets the matter out. It involves the square of the radius of the sphere, R^2. Beltrami took it, and investigated what it meant when the term R^2 was replaced by $-R^2$. He found that it made sense, but only inside a disc of radius R. Inside that disc one had, as it were, a map of something, but there was no longer any simple object of which it was a map. Inspired by Riemann, Beltrami decided that the map was enough: it was a complete intrinsic description of a surface upon which one could measure lengths. It was, therefore, a geometry, in the new sense of the term.

What did this new geometry look like? Geodesics appeared as straight lines, but distances were distorted: the same segment near the center of the disc was much shorter than a segment of the same Euclidean length but near the rim. In fact, lines were infinitely long according to the new formula for distances. Angles could also be defined in a way consis-

Figure 20. Geodetic Projection

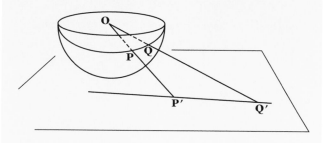

Box 5. The Beltrami disc

Points of non-Euclidean space appear as points inside the disc. Points on the boundary circle are not part of the map and do not represent anything. Non-Euclidean straight lines appear as straight lines in the disc. Several lines through the point P are shown. The line b meets the line a, and the line c is parallel to a — they meet on the boundary of the disc. The lines d and d' are among the lines through P that do not meet a.

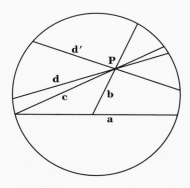

SAGGIO DI INTERPETRAZIONE DELLA GEOMETRIA NON-EUCLIDEA

DEL PROFESSOR

E. BELTRAMI

In questi ultimi tempi il pubblico matematico ha incominciato ad occuparsi di alcuni nuovi concetti i quali sembrano destinati , in caso che prevalgano, a mutare profondamente tutto l'ordito della classica geometria.

Questi concetti non sono di data recente. Il sommo *Gauss* li aveva abbracciati fino dai suoi primi passi nella carriera delle scienze , e benchè nessuno dei suoi scritti ne contenga l'esplicita esposizione, le sue lettere fanno fede della predilezione con cui li ha sempre coltivati e attestano la piena adesione che ha data alla dottrina di *Lobatschewsky*.

Siffatti tentativi di rinnovamento radicale dei principii si incontrano non di rado nella storia dello scibile. Oggi poi essi sono un portato naturale dello spirito critico

Beltrami's *Saggio*

tent with the new distance formula, and when this was done Beltrami found that the angle sum of a triangle was always less than two right angles. Finally, the curvature of the surface could be calculated, and it turned out to be a constant (equal to $-1/R^2$). The new geometry was non-Euclidean geometry, as Beltrami had wanted it to be.

The Beltrami disc (Box 5) certainly solves the problem once and for all, but it is not easy to use. The next step forward came when Felix Klein in Germany became interested in making sense of the sprawling profusion of geometrical work in the nineteenth century. In addition to Euclidean geometry, there were also, for example, studies of plane figures under various types of transformation that mapped lines to lines but changed lengths. These transformations are illustrated by the projection of one screen onto another by a point source of light; the geometry they give rise to is called projective geometry. Klein saw that the Beltrami disc fitted unexpectedly well into the projective way of thinking. The maps of the disc to itself that represent distance-preserving transformations in the sense of non-Euclidean geometry are, he showed, precisely the projective maps of the boundary circle to itself.

An even more useful new way of seeing the Beltrami disc occurred to the young French mathematician Henri Poincaré in June 1880, when he was changing buses as part of a field trip (he was a trainee mining engineer at the time). He had been studying a complicated network of triangles inside a disc, and to study them he had applied a transformation that straightened out the sides. This is how he recalled the event many years later, in 1908:

> At that moment I left Caen where I then lived, to take part in a geological expedition organized by the École des Mines. The circumstances of the journey made me forget my mathematical work; arrived at Coutances we boarded an omnibus for I don't know what journey. At the moment when I put my foot on the step the idea came to me, without anything in my previous thoughts having prepared me for it; that the transformations I had

Henri Poincaré

made use of . . . were identical with those of non-Euclidean geometry. I did not verify this, I did not have the time for it, since scarcely had I sat down in the bus than I resumed a conversation already begun, but I was entirely certain at once. On returning to Caen I verified the result at leisure to salve my conscience. (Poincaré [1908], 51–52.)

BOLYAI'S APPENDIX

Points of non-Euclidean space appear as points inside the disc. Points on the boundary circle are not part of the map and do not represent anything. Non-Euclidean straight lines appear either as straight lines in the disc perpendicular to the boundary circle, or as arcs of circles in the disc and perpendicular to the boundary circle. Several lines through the point P are shown. The line b meets the line a, and the line c is parallel to a – they meet on the boundary of the disc. The lines d and d' are among the lines through P that do not meet a.

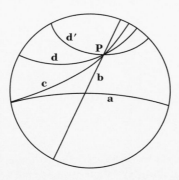

It is surely likely that as he boarded the bus he saw in his mind's eye the straight-sided triangles in the disc and recalled that he had learned that this was the Beltrami disc model of non-Euclidean geometry. Poincaré quickly drew two conclusions from this realization. One was that the complicated network of triangles was made up of congruent copies of the same non-Euclidean triangle, so the problem that had gen-

Figure 21. The (2, 3, 7) tessellation

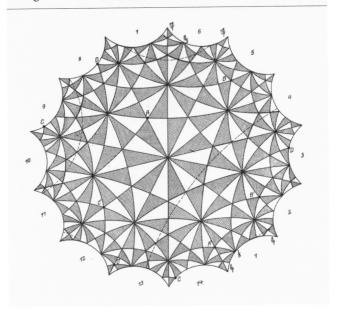

erated it could be studied geometrically. Another was that his original picture of triangles with curved sides actually had some advantages. The Poincaré disc model of non-Euclidean geometry is in fact the one generally in use today (Box 6).

The advantage of the Poincaré disc is that in it angles appear drawn to the correct size. So in Figure 21 all the triangles are congruent (in the sense of non-Euclidean geometry) and the angles are all either $\pi/2$, or $\pi/3$, or $\pi/7$.

To any one who accepted Riemann's framework for geometry, the Beltrami disc was conclusive proof that a geometry differing from Euclid's only in respect of the parallel postulate could exist. It permitted a mathematician to draw figures secure in the knowledge that they described something. The message spread quickly, and attempts to prove the parallel postulate shifted out of mainstream mathematics and into the hands of school teachers, engineers, and amateurs.

Teaching and Learning Geometry After 1870

AN EXCEPTIONALLY INTERESTING place to consider the reception of non-Euclidean geometry is Italy, which had been unified in the early 1860s. Several leading Italian mathematicians played a prominent role in the struggle and went on to occupy important positions in Italian political life. One of these was the distinguished geometer Luigi Cremona, who was decisive in ensuring that the new mathematics syllabus for the secondary schools gave prominent place to Euclid's *Elements*. A new edition was prepared by two more famous Italian geometers, Enrico Betti and Francesco Brioschi, who added many notes to assist the teachers, and thus Italian children were taught good criteria for discerning the true from the merely apparent by means of mental gymnastics (to paraphrase Cremona).

That this was not the only possible response is shown by the English example. In 1870 the Association for the Improvement of Geometrical Teaching

(AIGT) was founded in London to oppose the ritu-
alised, rote-learning of the *Elements* upon which too
many appointments in British life depended, and to
advocate the creation of a new, more flexible syllabus.
Even this proved difficult. Once the order of Euclid's
Elements is abandoned, the risk of a vicious circle
grows, and proofs sacrifice their logical and compelling character for seductive easiness. Books and
essays generated by the AIGT were criticized for their
lack of rigor, for example by the Oxford mathematician Charles Dodgson, better known as Lewis Car-
roll, in *Euclid and his Modern Rivals* (1879). In the
end the movement could not defeat the orthodoxy en-
shrined in the entrance examinations for Oxford and
Cambridge, and it faded away.

Both in Italy and in Britain, as elsewhere in Eu-
rope, the teaching of geometry was identified at the
school level with the teaching of Euclid's *Elements*,
and in particular with some form of the parallel postulate. In Italy, it was presented in Euclid's manner;
in Britain sometimes on the basis of arguments of an
allegedly more fundamental kind. But in each case
the message was generally underlined by a strong emphasis on the logical character of Euclidean geometry. The subject was not an empty mental exercise,
nor was it an empirical subject open to revision. It
contained truths of reason and exemplified logical
thought. An odd thing to assert after the discoveries
of Bolyai, Lobachevskii, and Gauss were finally becoming public knowledge.

One who was caught up most acutely in this dilemma was the French mathematician Jules Hoüel, a professor in Bordeaux. Throughout the 1860s he was trying to present elementary geometry in a clear and critical way, in which it would be obvious which propositions were drawn from experience and where logical argument took over. In the debate between a proponent of the AIGT and Cremona and Brioschi, Hoüel took the side of Euclid, at least as far as to deny that the *Elements* was illogical, whether or not it was suitable for use in schools. He preferred the *Elements* to Legendre's re-working of it because it had foundations that were easier to understand. Yet Hoüel was, as we have seen, an energetic translator of the new geometry, which he had begun to read about in January 1866, and it appealed to his critical spirit. To all those children in countries where education was the route to success, whether as engineers, doctors, lawyers, churchmen, philosophers, or teachers themselves, school education in mathematics offered a powerful picture of the subject strikingly at variance with what the experts themselves were coming to accept.

Such is the influence of education that many of these people opposed the idea of the new geometry with energy and some subtlety. Two examples must stand for many (and more can be found in Pont [1986]). In 1869 Jules Carton, a professor of mathematics at St. Omer, sent Joseph Bertrand what he claimed was a proof of the parallel postulate. His "proof" is typical of this unhappy genre: a complicated construction is produced which confuses the author and enough of his readers to allow him to deduce the conclusion to which he is already wedded. Rather more unusually, the expert to whom it was sent not only fell for the error but produced a simplified "proof" with an even more palpable flaw. And Bertrand was no mere expert. He was a professor at the École Polytechnique and the Collège de France, and a member of the prestigious Académie des Sciences in Paris. In 1874 he was elected the permanent secretary of the institution, and in due course became related by marriage to several leading French mathematicians – Charles Hermite, Emile Picard, and Paul Appell among them.

Carton's argument began by taking an arbitrary triangle ABC and joining to it a congruent copy BCA′, joining to it a further congruent copy A′BC′, and so on. He then counted angle sums in an erroneous way and deduced his conclusion. Bertrand proposed that any such chain of triangles can be enclosed in a quadrilateral. He then argued that the angular defect of the triangles in the chain can be made arbitrarily large by extending the triangles sufficiently, because each triangle has an angular defect of α say, and in non-Euclidean geometry $\alpha > 0$, so n triangles form a figure with an angular defect of $n.\alpha$, which grows with n. On the other hand, the angular defect of a figure enclosed in another one is greater in the larger figure. This follows by filling in the gap between the two figures with triangles and looking at the angles. The angular defect of the larger figure is the angular defect of the smaller one plus the sum of the angular defects of the triangles in the gap. But this means that the angular defect of a chain of triangles cannot be more than 4π, contradicting the previous conclusion. The mistake in the argument is the assumption that any figure can be contained in a quadrilateral (it is an enjoyable, and not completely trivial exercise to show that this cannot be true).

Bertrand, however, presented Carton's argument to the Académie des Sciences in Paris on December 20, 1869 (Bertrand, *Comptes Rendus* [1869], pp. 1265–1269). To make it clear that he had not merely nodded he then published a short note with his simplification of Carton's proof (Bertrand, *Comptes Rendus* [1870], pp. 17–20). Darboux, Hoüel, and Beltrami, who were very active in bringing non-Euclidean geometry to France at the time were, of course, appalled, and others were drawn in. Newspapers took up the affair, and when it was demonstrated publicly not only that Carton's supposed proof was not new (an Italian mathematician, Minarelli, had published it in the *Nouvelles Annales* 8 [1849], p. 312), but that it was, of course, fallacious, Bertrand finally withdrew his support.

The second example is more disturbing and more notorious. Among the many attempts by philosophers to show that after all geometry must be Euclidean, Gottlob Frege's stand in the sharpest contrast to the rest of his work. Whatever the merits of his contributions to the philosophy of mathematics, which continue to generate discussion to this day, one foot remains mired in the clay of Euclidean geometry, which Frege continued to defend all his life. As he put it: "No one can serve two masters. One cannot serve truth and untruth. If Euclidean geometry is true, non-Euclidean geometry is false, and if non-Euclidean geometry is true, Euclidean geometry is false." (Frege, *Nachgelassene Schriften* [1969], pp. 183–4; the manuscript was written between 1899 and 1906.) But he did not mean by this to hold both systems in abeyance until some way of choosing between them could be found. Rather, he thought that on philosophical grounds the choice was already made, in favor of Euclidean geometry. The comparison is between science and alchemy, science and astrology, and now between Euclidean and non-Euclidean geometry. Euclidean geometry is two thousand years old – can it really be doubted? If not, concluded Frege, non-Euclidean geometry must be counted among the non-sciences.

Not only does this passage show a remarkable degree of confidence in Euclidean geometry, it makes clear that Frege just could not see any value in a hypothetical mathematical system. It was a case of, as he put it, true or false, in or out. The most remarkable aspect of the whole position, however, is that it was to turn out that the two geometries, Euclidean and non-Euclidean, stood or fell together. They were relatively consistent, as we shall see below. Nothing could more sharply indicate how badly Frege failed to understand the new geometry than his aggressive insistence on a dichotomy that does not itself exist.

A philosopher of Frege's calibre cannot plead that he was the prisoner of his education. More excusably, the minds of many who would never have considered themselves mathematicians were filled with the eternal verities of geometry, conveyed by school teachers who cannot all have understood them and were mostly unaware of the disputes and alternatives that crowded the final decades of the nineteenth century. It is the experience of being taught, not the experience of life or of real lines, that persuaded most people that the parallel postulate was true. Being told that it is so, and finding that instruction thereafter proceeds on that assumption (whether smoothly or not) was the process whereby Euclidean geometry was taught to any one who acquired an advanced education at all. The result was the existence of a large number of people who either thought that they knew for certain that the parallel postulate was correct, and could not be persuaded otherwise, or who were prepared to be shocked and excited on hearing that it was not. Poincaré's popular essays, which are discussed below, tapped a real nerve, and reached as far as such modern painters as Francis Picabia and Marcel Duchamp (as Linda Henderson excellently describes in her 1983 book). As the nineteenth century drew to a close, the public knew very well that the experts were changing their opinions about the nature of geometry.

Taking Stock

A NEW GEOMETRY is a momentous thing to have discovered, and it took some time for its implications to sink in. It was easier for mathematicians to employ it than to engage with it. When they did it was clear that the philosophical consequences were profound. There cannot, after all, be two incompatible accounts of physical space that are both true. It follows that one of the geometries must be false (and perhaps both, but this radical view was never espoused). The question then arose: which one? It was obvious that space was Euclidean to all intents and purposes. But the empirical question hung in the air: might not some delicate observations decide the matter one way or another?

The most original answer to this conundrum was offered by Poincaré and goes under the name of conventionalism. In an essay published in 1895, and reprinted in *La Science et l'hypothèse* as "L'Espace et la Géométrie" in 1902, he surveyed non-Euclidean geometry, which he described as that of a spherical world in which temperature falls off as one moves away from the center according to a prescribed rule (which he gave). Inhabitants of this world will find it infinite, and if thermal equilibrium is always achieved instantly as they move around they will have no sense of temperature varying. They will not think that they and their measuring instruments expand and contract as they move about, but that they are constant.

Figure 22. Escher, Circle Limit IV

An impression of this world is conveyed by a famous M. C. Escher print, suggested to the artist by the mathematician H. S. M. Coxeter (Figure 22). All the white "angels" are the same size, and all the black "devils" are. They appear to contract as they move outwards, but intrinsically they do not. The previous figure (Figure 21, which is due to Poincaré's contemporary Felix Klein in 1880) shows a similar configuration of black and white triangles, and shows clearly that the straight lines of the figure occur as diameters of the circle or arcs of circles perpendicular to the boundary circle. Remarkably, Klein did not appreciate that this picture was a picture of the non-Euclidean disc; perhaps he was misled by the fact that in his model non-Euclidean straight lines appear straight.

What geometry will the inhabitants of the cooled sphere ascribe to the space they live in? What, for example, will they say is a straight line? According to the rule Poincaré gave, they will say that the spherical world is described by non-Euclidean geometry, and their straight lines will be diameters of the sphere and arcs of circles perpendicular to the boundary of the sphere.

The conclusion Poincaré drew from this thought experiment is that while we and the inhabitants of this spherical world disagree about the nature of their world, neither of us is right. All any experiment could do would be to confront us with a dilemma. Suppose the physical embodiment of four straight lines presented us with a figure (a "square") having four equal sides and four equal angles, but the angles were each less than a right angle. Should we conclude that the geometry of space is non-Euclidean, or that geometry is Euclidean but our way of creating lines is deficient? The inhabitants of Poincaré's spherical world could do either, and are not forced by logic alone to choose which. We ourselves, knowing since Einstein that light rays are curved by large gravitational objects (such as the Sun) might draw a "square" of light rays connecting four equally spaced points on the

Earth's orbit, and find that the angles were each less than a right angle. We can conclude either that space is Euclidean but light rays are curved, or that light rays are straight and space is non-Euclidean (but according to Poincaré we cannot say which. All we can do is prefer one on the grounds that it is more convenient, and Poincaré presumed that we would always prefer Euclidean geometry. This act of choice establishes a convention, and like all conventions it is neither true nor false.

Poincaré's grounds for making this choice are unexpected. It is not that Euclidean geometry takes place in a flat space that makes him prefer it. Rather, it is that we learn geometry as children, and did as a species, by moving rigid bodies around. We learn that this motion of a rigid body followed by that one is equivalent to a third motion, and indeed we learn that the set of all possible motions of a rigid body form a mathematical object dear to Poincaré's heart, that of a group. Now Euclidean geometry and non-Euclidean geometry have different groups, and one of these is indeed simpler on mathematical grounds. This is the group of Euclidean motions, and it is simpler because one can disentangle translations and rotations in it. It is on these grounds that Poincaré found Euclidean geometry simpler.

Poincaré's argument may well not sound very convincing. The point is rather that it was firmly grounded in the psychological speculations of its day. For Poincaré what mattered was to explain how the human mind came to have ideas at all. His account was Kantian, in giving priority to such a question, and post-Kantian in attaching such emphasis to psychology. But it was resolutely humanist; Poincaré set his face against a trend to make mathematics a purely logical activity which, as we shall see, was also inspired by the discovery of non-Euclidean geometry.

Nor indeed did his conventionalism find much favor outside his native France, where Poincaré enjoyed the status of a national sage. Federigo Enriques in Italy, who was an eminent geometer and close to Poincaré on many positions, felt that we could always distinguish physical properties from geometrical ones, and so the thought experiment was flawed. Since Einstein's discovery of general relativity theory, which proposed that space was inhomogeneous and curved by bodies, most writers have agreed with Enriques.

Poincaré did, however, resolve another question that was left over from the work of Beltrami and others. They had established that, on certain assumptions, there was a self-consistent treatment of non-Euclidean geometry. No-one doubted that there was a self-consistent treatment of Euclidean geometry. But how are the two geometries related? Poincaré observed in his essay "Les Géométries non euclidiennes," also reprinted in *La Science et l'hypothèse*, that if there was a self-contradiction in non-Euclidean geometry, it would result in a self-contradictory figure in non-Euclidean geometry. This would be an impossible figure in the Poincaré disc. But any such figure is also a figure in Euclidean geometry, and being impossible it implies a contradiction in Euclidean geometry. The surprising conclusion is that Euclidean and non-Euclidean geometry are relatively consistent: either they are both logically possible, or neither is. It is a novel twist to the story that had any one discovered a valid proof that non-Euclidean geometry is impossible, they would also have shown that Euclidean geometry is also impossible – not what they were looking for at all!

Logical Worries

BUT HOWEVER the empirical questions were to be resolved, the mere fact that there was a significant empirical question about geometry was an embarrassment. Geometry, Euclidean geometry, had been the paradigm of truth, a rich system of deductions about the world obtained by thought alone. As such it was a secure base for any more obviously empirical pursuit, such as mechanics or any branch of physics. Now there was no adequate store of results. Bolyai's Absolute Geometry was a small collection of theorems, and almost all the useful results in geometry now stood in conflicting versions, Euclidean and non-Euclidean. It was not just that thought alone seemed impoverished; the very fact that mathematics had been wrong for so long had to be confronted. Mathematicians began to ask: just how good was the rest of the *Elements*? What, for example, is a plane? What is a straight line, and what is it about them that makes them different from the hyperbola and its asymptote?

Euclid's definitions do not help. In Heath's edition of the *Elements* a straight line is defined as a line lying evenly with the points on itself, and a plane surface as the surface which lies evenly with the straight lines on itself. Recently, Lucio Russo has questioned the authenticity of these definitions. He suggests that they have been taken from Heron of Alexandria's *Definitions*, where they occur in amplified form, and interpolated by a writer in late antiquity. Whatever view one takes of the matter, it is generally agreed that these and the other opening seven definitions of Euclid's *Elements* differ in character from the rest in being non-technical and, which is far worse, useless. They are at best vague gestures towards a concept already understood by speaker and hearer, writer and reader, and they are quite incapable of resolving any serious disagreement. It would have been better for Euclid to have accepted Aristotle's dictum that one has to start with some undefined terms (as perhaps he did).

A useful way to approach tricky questions of definition is to ask what properties of the plane are being used when theorems are proved. For example, it is said that if two points on a line lie in a given plane then the entire line lies in that plane. Some ancient writers, such as Heron of Alexandria, and some modern ones, such as Robert Simson, who edited an edition of Euclid's *Elements* in 1756, and Gauss's teacher Kästner took this as a definition of plane. Interestingly enough, in an 1829 letter to his friend, the astronomer F. W. Bessel, Gauss had criticized this view on the grounds that it said too much: a plane could be defined more simply and this statement about it was really a theorem.

Gauss wrote an account of what he might have in mind at the end of a book by J. W. Lehmann, published in 1829. He wished to define a plane as a "surface generated by rotating a straight line around an axis with which it formed right angles." It is necessary to show that this definition does not involve the choice of axis in any essential way; in other words, that if a different axis is chosen then either the planes are parallel, or they meet in a line, or they coincide. Gauss did not spell out how this can be done, but he was only writing for himself, not publication. He did, however, go on to sketch a proof that on this definition if two points lie in a plane then the entire straight line joining them lies in the same plane.

Gauss's arguments, and those of two contemporaries, August Crelle and H: W. Deahna, raised another issue which they sometimes confused with the one just mentioned. They asked if this intersection property was obvious: if three points A, B, and C lie in a plane and are joined by the line segments AB, BC, and CA, and if the line DE passes between the points A and B then it passes through either a point of the segment *BC* or the segment *CA*. Crelle, who was a distinguished railway engineer and the founder of what rapidly became the leading journal for mathematics in Germany, was muddled on this issue, but Gauss read Deahna's doctoral thesis on the subject and found it satisfactory.

Another topic where Euclid's *Elements* was found wanting occurs in its very first proposition: Euclid assumed that two circles with their centers at two points *A* and *B* and radii *AB* must meet. Ancient commentators observed that it is only in Book 3 of the *Elements* that it is shown that two circles cannot meet in more than two points without coinciding entirely, so it should not be assumed here, but they skipped over the need to stipulate that there are any common points at all.

As the nineteenth century wore on, these glitches in the *Elements* were more and more discussed. Moritz Pasch singled out the intersection property and made it an explicit assumption in his 1882 lectures on geometry; it is generally called Pasch's axiom to this day. He, and commentators like Wilhelm Killing and Bertrand Russell, noted the missing assumption about circles. Other commentators worried about the reliance on diagrams in Euclid's *Elements*. It is almost impossible to follow the proofs in the *Elements* without drawing figures, and the growing worry was that the proofs were only convincing because the diagrams slipped something past the reader.

To indicate the problem, the Cambridge mathe-

Figure 23. Every triangle is isosceles!

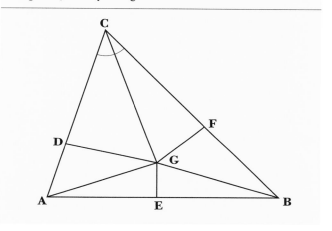

maticlan W. W. Rouse Ball devoted a whole chapter of his book of 1892 to fallacious arguments driven by convincing, but flawed, diagrams. Here, for example, is his famous "proof" that every triangle is isosceles.

> Let *ABC* be any triangle [see Figure 23]. Draw the internal bisector of angle *C* and the perpendicular bisector of side *AB*, and let them meet at *G*. Drop perpendiculars *GD* from *G* to *AC* and *GF* from *G* to *BC*. Draw *AG* and *BG*.
>
> We are going to show that the sides *AC* and *BC* are equal.
>
> Triangles *CGD* and *CGF* are congruent because their angles are equal in pairs and they have a side in common. So the lengths *GD* and *GF* are the same, and so are the lengths *CD* and *CF*.

Triangles *GEA* and *GEB* are congruent, because *GE* is the perpendicular bisector of *AB*. So the lengths *GA* and *GB* are the same.

Finally, triangles *GDA* and *GFB* are congruent, because they are both right-angled, and we have just shown that two of the sides are equal in pairs. So the lengths *DA* and *FB* are the same.

Adding the results *CD* = *CF* and *DA* = *FB*, we find that *CA* = *CB*, as claimed.

Something is badly wrong. Perhaps it is the diagram. We have drawn it with *G* inside the triangle. Perhaps it lies on the edge *AB*. The reader can quickly run through the argument in this case, and reach the same alarming conclusion. The same is true if it is supposed that *G* lies outside the triangle, or even if the case is considered where the internal bisector of angle *C* and the perpendicular bisector of side *AB* are parallel. It would seem that geometry shows that every triangle is isosceles, and therefore that any two lengths are equal!

It is an interesting exercise to find the flaw, and the answer is accordingly deferred to the footnotes.[21] The intellectual point that geometric diagrams can mislead should, however, by now be established.

The stage was set for new foundations for geometry, ones which would be immune to the criticisms that were piling up against the venerable *Elements*. The first to make a thorough-going attempt was Pasch, who published his book, *Vorlesungen über neuere*

Geometrie in 1882. He rejected turning geometry into coordinate geometry, on the grounds that proofs in coordinate geometry were heavily arithmetical, but the concepts of geometry were only occasionally so. Instead he endeavoured to state all the undefined or primitive concepts necessary for plane projective geometry (which is logically simpler than Euclidean geometry) and to formulate precisely the minimal number of facts about these concepts necessary to get geometry going. For example, the first basic fact was that there is always a unique segment joining any two points. He claimed that these basic facts were grounded in observation, citing Hermann van Helmholtz's paper "On the origin and significance of the geometrical axioms" in his support. Then he set about deducing what he could from these basic facts purely logically and without any further appeal to intuition. Eight basic facts gave him a geometry of line segments, four more (including the axiom now named after him) were needed to generate all of plane geometry. He dealt with the idea of congruence in the same way: intuition gave him the basic facts, and logical deduction thereafter the required results. His intuition was such that he could in fact coordinatise the plane; later writers were more parsimonious.

In the 1890s in Italy, Italian mathematicians were also engaged on a similar pursuit. They too took projective geometry but were less concerned to refine the intuitions of ordinary experience, more to study the axioms in their own right. Gino Fano observed that while it is fundamental in plane projective geometry that any two points determine a unique line and any two lines meet in a unique point, it is not necessary that the plane be at all familiar. It is possible that it consist of only finitely many points, say seven, or fifteen. He showed that such planes have unfamiliar properties, but are not logically absurd. In his "Sui fondamenti della geometria" Giuseppe Peano showed that Desargues's Theorem is automatic in projective spaces of dimensions three or more, but gave no proof in the plane case, which suggested that something unexpected might be possible.

A third Italian, Mario Pieri, took the decisive step of breaking with Pasch, and abandoned completely any intention of formalizing intuitions based on experience. Instead, as he wrote in his "Sui principi che reggiono la geometria di posizione" of 1895, projective geometry is treated "in a purely deductive and abstract manner . . . independent of any physical interpretation of the premises," and primitive terms, such as line segments, "can be given any significance whatever, provided they are in harmony with the postulates which will be successively introduced." In his presentation of plane projective geometry of 1898, the *Principii della Geometria di Posizione*, he put forward nineteen axioms (typically: any two lines meet). The idea that geometry should be studied entirely rigorously and with no appeal to intuition, which had become something at best redundant and at worst dangerous, was now abroad.

Unluckily for the Italians, the idea was propagated with much more success by the leading German mathematician of the day, David Hilbert. In his *Grundlagen* (1899), he divided up the axioms of his presentation of geometry into five kinds, according to the kind of intuition they formalized. He then studied the systems of geometry generated by interesting subsets of these axioms. To check that the axioms were independent he also found consistent sets of axioms consisting of some of his original list with the negation of others; much non-Euclidean geometry has the same axioms as Euclidean geometry except for the parallel postulate (which is replaced by its negation). In particular, he found that the axioms of plane projective geometry are consistent with the negation of Desargues's Theorem, which entails the shocking result that Desargues's Theorem may be false in the plane. It follows that there are projective planes that cannot be embedded in a projective three-dimensional space. Hilbert's original proof was very complicated, and was so greatly simplified by the American astronomer Forest Moulton that Hilbert replaced his argument with Moulton's in later editions of his work.

In 1900 the first International Congress of Philosophers met in Paris. This was followed back-to-back by the second International Congress of Mathematicians – evidence of the growing internationalism of intellectual life. The philosophers heard many papers on geometry, and the mathematicians heard yet another Italian, Alessandro Padoa, on a new system of definitions for Euclidean geometry, which drew on ideas of Pieri and others. Hilbert took the occasion to propose twenty-three problems in mathematics, as part of an eye-catching attempt to predict the future of mathematics. Several of these were in the spirit of his account of geometry, so however little Hilbert might have known of Italian work when he began, and the evidence is slight, he surely knew of it by now.

The Italians compounded their relative obscurity with a desire to write mathematics in a heavily symbolic style that sought to exclude natural language, so deep did their distrust of intuition go. This made them difficult to read. Hilbert, on the other hand, was the leading mathematician at Göttingen, then the most formidable institution in mathematics in the world, where he was the center of a group of bright ambitious young men. He realised that his approach to the study of axioms enabled him to prove a new kind of theorem. These were theorems about theorems: such as, that, in the presence of certain more elementary axioms, Desargues's Theorem can be false in geometries of two dimensions, but not of three. As a friend wrote to him, congratulating him for creating a branch of mathematics much broader than geometry: "You have opened up an immeasurable field of mathematical investigation which can be called the "mathematics of axioms" and which goes far beyond the domain of geometry." (quoted in Toepell [1986] p. 257).

Hilbert's work was also clearer about what systems of axioms should try to be. For him, the most logical starting point was a set of axioms which were the most efficacious and easy to apply; they might well not be intuitively obvious at all. This is far from the writing down of axioms and the drawing of conclusions. As David Rowe has argued (see his introductory essay to Hilbert [1992]) for Hilbert mathematics has two tasks: to discover systems of relations and their logical consequences; and to provide a fixed structure with the simplest possible foundations – this is the task of axiomatics. An active stand in twentieth-century mathematics was to pursue these broader implications.

Poincaré was favorably impressed by Hilbert's *Grundlagen* of 1899, which he reviewed in 1902. He liked the novelty of Hilbert's alternatives, but he noted that it was now impossible to give a psychological account of geometry, and he lamented its enforced absence. But where Poincaré differed most starkly from Hilbert is in the nature of axioms. For Poincaré, the most logical starting point for mathematics was in the simple intuitions of the mind about number and shape; conclusions spread out from there in, perhaps, ever more elaborate ways. We see again how Poincaré's psychologism colored his ideas about all of mathematics.

The Nature of Space

AFTER 1905, when Einstein had published his theory of special relativity, he turned his mind to the nature of gravitation. In 1907 he was looking at the case of the geometry on a rotating disc. Special relativity theory predicts that if an object is moving with respect to another, it will appear contracted along its direction of motion. So to an observer at the center of the disc, a meter stick out towards the boundary will appear contracted. This contraction will be greatest along the direction at right angles to the radius joining the observer to an end of the stick. The situation is very similar to the cooled sphere Poincaré had described (and Einstein had studied Poincaré's popular essays with great care at university). If the disc is such that its outer edge is rotating at the speed of light, Einstein showed that the observer at the center will indeed think that the geometry on it is a non-Euclidean geometry.

Einstein's rotating disc is another thought experiment. He never seriously maintained that space might be rotating. But when he published the equations that describe how space is curved by matter and gravitation therefore exerts its effect, it was noted that for two-dimensional space there is a solution which resembles a paraboloid. It is called Flamm's paraboloid after its discoverer (Figure 24). The paraboloid is everywhere negatively curved, but it gets less curved as one moves outwards. In the same way,

Figure 24. Flamm's paraboloid

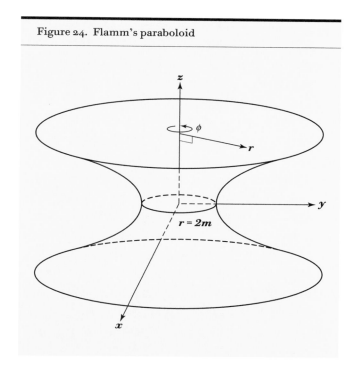

the gravitational effect of a mass diminishes with distance. Flamm's paraboloid shows that our part of space may indeed be taken to be non-Euclidean, but not exactly as Bolyai might have thought. Instead it is markedly non-Euclidean near the Sun, and less so as one moves outwards. But it is, in any case, not Euclidean.

Conclusions

WE HAVE COME A LONG WAY from the certainties of the practical man and woman, the scientist, the mathematician, the educator, and the philosopher of the seventeenth and eighteenth centuries. Euclidean geometry is no longer regarded as true. It is not the geometry of physical space, and it is not preferable on any logical grounds. It has a certain approximate accuracy, and it is much easier than its rival; practical people naturally and sensibly continue with it. But the scientists, mathematicians, and the philosophers have moved on. Scientists have Einstein's theory of general relativity. Mathematicians no longer find Euclidean geometry the epitome of rigor, nor do they even accept that geometry must be cast in anything like Euclidean terms. Educators may not have embraced non-Euclidean geometry, but they despaired of plugging the holes in Euclid's *Elements* and may even be nervous of the coordinate geometry. Philosophers too must accept the plurality of modern geometry and physics. It can all look like a defeat for Euclid.

And yet, and yet. It is remarkable that the Greeks singled out the parallel postulate, and generally did as well as they did. There are flaws in Euclid's *Elements*, but assuming the truth of the parallel postulate was not one of them. After two thousand years it turned out that making it an assumption, for whatever reason, was correct. It was those who offered proofs of the postulate who were mistaken. And by making it clear, over an expanse of thirteen detailed books, what followed from what, Euclid set a style that, revitalized by the Italians and by Hilbert, in the even more austere form of abstract axiomatics spread to much of mathematics. In this light, what Bolyai, Gauss, and Lobachevskii did was to recapture the spirit of Euclid's *Elements*. When Bolyai created a whole new world out of nothing, he thereby liberated mathematics from being solely the servant of science, and this in turn enabled mathematics to make fresh contributions to the study of the natural world.

Notes to the Introduction

1. Newton. *Principa,* 408-09.
2. Newton. *Principa*, 943.
3. Pont. *L'aventure des parallèles*, 231-33.
4. Kant. *Philosophical Correspondence, 1759–1799*, number 61.
5. Gauss. *Werke*, VIII, 159.
6. Gauss. *Werke*, VIII, 168-69.
7. Gauss. *Werke*, VIII, 177.
8. Gauss. *Werke*, VIII, 200.
9. Gauss. *Werke*, VIII, 201.
10. Gauss. *Werke*, VIII, 175-76.
11. Gauss. *Werke*, VIII, 180-81.
12. Gauss. *Werke*, VIII, 181.
13. Gauss. *Werke*, VIII, 186-88.
14. Stäckel. *Wolfgang und Johann Bolyai,* 81.
15. Stäckel. *Wolfgang und Johann Bolyai,* 81.
16. Stäckel. *Wolfgang und Johann Bolyai,* 81.
17. Stäckel. *Wolfgang und Johann Bolyai,* 81.
18. Gauss. *Werke*, VIII, 220-24.
19. Biographical information is taken from Lobachevskii, *Zwei geometrische Abhandlungen*.
20. Sartorius. *Gauss zum Gedächtnis*, 82.
21. The mistake is to assume that D and F, the feet of the perpendiculars from G to AC and from G to BC both lie on the triangle. In fact, if the edge CA is shorter than edge CB then D lies outside CA while F lies between C and B. The consequence is that the final deduction is flawed, and the edge lengths of the triangles are obtained by an addition and a subtraction.

Legendre's mistake (see p. 40)

When Legendre's construction is carried out repeatedly in the Poincaré disc, it becomes clear that the sequence of points A, A′, A″ . . . , $A^{(n)}$ . . . eventually enters a region with the property that no line through a point $A^{(n)}$ in the region meets both AB and AC, however far they are extended. So the angular defect cannot be doubled indefinitely, and Legendre's argument collapses.

BOLYAI'S "SECTION 32"

I N SECTION 32 of the Appendix, János Bolyai
made a series of observations whose significance
seems to have escaped him and most of his commentators, even though he prefaced them with the
remark that he here showed the method of resolving
problems in the new geometry, "which being accomplished (through the more obvious examples), finally
will be candidly said what this theory shows." To appreciate what he then did, it is best to take a step back
and recall what would have been well known to all his
mathematical readers.

Bolyai considered a curve in the plane given by
an equation of the form $y = f(x)$ with respect to the
usual system of Cartesian (x, y) coordinates. Questions about the area of the region bounded by the
curve, two vertical lines, and the x-axis are answered
by integration, a fundamental process of the calculus. The calculus also allows mathematicians to answer questions about how far it is along the curve from
one point to the next. The same is true of areas and

lengths on a surface, whenever that surface has been mapped onto a plane. If, for example, the earth's surface is represented on a flat, Euclidean, plane, there are formulae for the area of a region of the sphere in terms of latitude and longitude, and for the length of a curve on the sphere, which his readers might well have known. A glance at Mercator's projection shows that the formulae cannot be as simple as they are in ordinary plane geometry: equal increments of y on the map clearly correspond (as one goes North) to smaller and smaller steps on the surface of the Earth. So the formulae are different, but the idea is the same.

Without as much as a hint in the direction just outlined, Bolyai supposed that his readers would recognise these arguments and appreciate that they will work in the altered setting of his new geometry. He drew a picture of a curve ABG in the familiar Cartesian plane with x- and y- axes and outlined an interpretation of it as a picture of non-Euclidean geometry drawn in a Euclidean plane. He supposed that the x-axis represented a non-Euclidean straight line, took two points C and F on the x-axis, and drew perpendiculars to the x-axis through C and F, meeting a curve at points B and G respectively. Through the point B he then drew the ray BH parallel to the x-axis, making equal angles FCB and HBC. (See Figure 25.)

This is not an impossible picture, but it requires an interpretation that Bolyai was unwilling to provide. The perpendiculars CB and FG must be what he elsewhere called L-lines. It is not at all obvious that

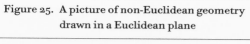

Figure 25. A picture of non-Euclidean geometry
drawn in a Euclidean plane

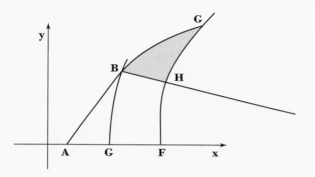

they can be represented as Euclidean straight lines.
Let us grant that they can. Bolyai then asked his readers to consider what happens to the triangle *BGH*
(shown shaded in the figure) when the distance *CF*
becomes very small. The arc-length *BG* along the
curve becomes very small, and better and better ap-
proximated by the non-Euclidean length of the non-
Euclidean line segment *BG*. As in Euclidean plane
geometry, knowledge of very small lengths will yield
knowledge of arbitrary finite lengths along the curve
by means of integration. Now, the formula connecting the sides of the very small triangle *BGH* is obtained from the formulae for an arbitrary finite non-
Euclidean triangle, and this is exactly what Bolyai had
established in the previous paragraph. Note, though,
that any finite non-Euclidean triangle *BGH* differs
from the actual shaded triangle Bolyai was consider-

ing in two respects. Not only is the side BG part of an arc, but the side HG is part of an L-line, not a non-Euclidean straight line. The passage from an arbitrary finite but very small triangle to an approximation whose sides are all non-Euclidean straight lines is therefore not an elementary one.

Bolyai escaped the pedagogic problem, not for the first or only time in the *Appendix* by saying: "It can be demonstrated" that the requisite approximation is $BG^2 = dy^2 + BH^2$. He then used this formula to deduce formulae for the areas and volumes of some non-Euclidean figures.

As a matter of fact, Bolyai was right. His approximation formula can be shown to be correct. It is perhaps a pity that he did not show how to prove it. But more to the point, he missed a great opportunity to start again and provide an argument in support of non-Euclidean geometry that would have had a better chance of convincing the skeptical reader.

Bolyai could have started anew, and said that he would draw pictures in the usual (x, y)-plane in which very small distances were to be measured according to his new rule, and arbitrary finite distances determined from these by the usual process of integration. With a bit of extra work, he could have shown that the entire picture of non-Euclidean two-dimensional geometry could appear in the right half-plane (the region defined by $x > 0$), and that in his new space straight lines were curves of a certain appearance, that two lines were parallel if (in a typical case) they met on the *y*-axis, and that the trigonometric formulae for triangles were those he had derived earlier. Equal non-Euclidean distances would appear to shrink as one neared the *y*-axis (the exact opposite of the odd feature of Mercator's projection of the sphere). All the same results, in other words, but derived from a much more plausible starting point.

This is not to say that the world would have been convinced. The obstacles to accepting the new geometry were surely too great. Indeed, the conceptual leap needed to formulate the new approach is precisely what we consider one of Riemann's great contributions to have been. But the fact that Bolyai got as close as he did to formulating the elements of his new geometry in terms of the calculus is striking testimony to his insight, and seems not to have been appreciated sufficiently in his day or since.

A PORTRAIT AT LAST

———

I N J U L Y 2 0 0 2 the Hungarian Academy of Sciences celebrated the 200th anniversary of the birth of János Bolyai. There they announced that for the first time it seems there is a reliable picture of him, which deserves to replace the portrait of unknown provenance and doubtful authenticity that is sometimes shown. The new picture comes from a contemporary relief on the façade of the Palace of Culture in Marosvásárhely, which carries the name of János Bolyai and was apparently identified by János's son and others at the time. The other five figures are known to be those of János's contemporaries. Plenary lecturers were given a handsome medallion 6.5 cms in diameter made by Kinga Széchenyi and based on the new portrait, a copy of which is shown opposite.

BIBLIOGRAPHY

Aristotle. *De Caelo* [On the Heavens]. Ed. and Trans. J. L. Stocks. In vol. 2 (published 1930) of *The Works of Aristotle Translated into English.* Ed. Sir W. D. Ross. Oxford: Oxford University Press, 1928–1952.

Ball, W. W. Rouse. *Mathematical Recreations and Essays.* Cambridge: Cambridge University Press, 1892.

Beltrami, Eugenio. "Saggio di interpretazione della geometria non-Euclidea," *Giornale di matematiche* 6 (1868): 284–312.

Bolyai, Farkas. *Tentamen juventutem studiosam in elementa matheseos purae, elementaris ac sublimioris, methodo intuitiva, evidentiaque huic propria, introducendi: cum appendice triplici . . .* Maros Vásárhelyini: Typis Collegii Reformatorum per Josephum et Simeonem Kali de felső Vist., 1832–1833.

Bolyai, János. *Appendix: Scientiam spatii absolute veram exhibens: a veritate aut falsitate Axiomatis*

*XI Euclidei (a priori haud unquam decidenda) in-
dependentem; adjecta ad casum falsitatis, quadratura circuli geometrica*. Maros Vásárhelyini:
Typis Collegii Reformatorum per Josephum et
Simeonem Kali de felső Vist., 1832–1833.

Bolyai, János. "La Science absolue de l'espace,"
(Trans. Guillaume Jules Hoüel) *Mémoires de la
Société des Sciences physiques et naturelles de Bor-
deaux* 5 (1867): 189–248.

Bolyai, János. "Sulla scienza della spazio assoluta-
mente vera," (Trans. Giuseppe Battaglini) *Giornale di matematiche* 6 (1868): 97–115.

Bolyai, János. *The Science Absolute of Space: Independent of the Truth or Falsity of Euclid's Axiom
XI (Which Can Never be Decided a priori)*. Trans.
George Bruce Halsted. 4th ed. Austin: The University of Texas, 1896.

Bonola, Roberto. *Non-Euclidean Geometry; a critical
and historical study of its developments. Authorized
English translation with additional appendices by
H. S. Carslaw. With an introd. by Federigo Enriques.
With a suppl. containing the George Bruce Halsted
translations of The science of absolute space, by John
Bolyai [and] The theory of parallels, by Nicholas
Lobachevski*. Chicago: Open Court, 1912. Repr.
New York: Dover Publications. 1955.

Bonola, Roberto. *La geometria non-Euclidea: esposizione storico-critica del suo sviluppo*. Bologna:
Zanichelli, 1906.

Byrne, Oliver. *The First Six Books of the Elements of Euclid in which Coloured Diagrams and Symbols are Used Instead of Letters for the Greater Ease of Learners*. London: William Pickering, 1847.

Clairaut, Alexis-Caude. *Éléments de Géométrie*. Paris: Chez Lambert & Durand, libraires, 1741.

Descartes, René. *The Geometry of René Descartes*. Trans. David Eugene Smith and Marcia L. Latham. Chicago; London: The Open Court Publishing Company, 1925.

Dodgson, Charles L. *Euclid and his Modern Rivals*. London: Macmillan and Co., 1879.

Engel, Friedrich and Paul Stäckel. *Die Theorie der Parallellinien von Euklid bis auf Gauss: eine Urkundensammlung zur Vorgeschichte der nicht-euklidischen Geometrie in gemeinschaft mit Friedrich Engel*. Leipzig: B. G. Teubner, 1895.

Includes a reprint of Johann Heinrich Lambert's "Theorie der Parallellinien." Originally published in *Leipziger Magazin für die reine und angewandte Mathematik* (1786): 139–64; 325–58.

Enriques, Federigo. *Problems of Science*. Trans. Katharine Royce. Chicago: Open Court, 1914.

Translation of *Problemi della scienza* (Bologna, 1906).

Euclid. *The Thirteen Books of Euclid's Elements*. Ed. and Trans. Sir Thomas L. Heath. 3 vols. Cambridge: Cambridge University Press, 1926. Repr. New York: Dover, 1956.

Euler, Leonhard. *Elements of Algebra*. Trans. John
     Hewlett. Ed. C. Truesdell. New York; Heidelberg:
     Springer Verlag, 1984.

> Facsimile of the 1840 London edition. Orig-
> inally published as *Vollständige Anleitung zur Al-
> gebra* (St. Petersburg, 1770).

Fano, Gino. "Sui postulati fondamentali della geome-
     tria proiettiva," *Giornale di matematiche* 30
     (1892): 106–131.

Frege, Gottlob. *Nachgelassene Schriften und Wis-
     senschaftlicher Briefwechsel*. Ed. Hans Hermes,
     Friedrich Kambartel, and Friedrich Kaulbach.
     Hamburg: Felix Meiner Verlag, 1969, 1976.

> Volume 1: *Nachgelassene Schriften*

Gauss, Carl Friedrich. *Werke*. Leipzig: K. Gesellschaft
     der Wissenschaften zu Göttingen, 1870–1933.

> Vol. 8. Arithmetik und Algebra. Nachträge zu
> Band I-III. Leipzig: B. G. Teubner, 1900.

Gray, Jeremy J. *Ideas of Space, Euclidean, Non-Eu-
     clidean and Relativistic*. 2nd edition. Oxford: Ox-
     ford University Press, 1989.

Helmholtz, Hermann von. *Epistemological Writings:
     The Paul Hertz/Moritz Schlick Centenary Edition
     of 1921 with Notes and Commentary by the Edi-
     tors*. Trans. Malcolm F. Lowe. Ed. Robert S. Co-
     hen and Yehuda Elkana. Boston Studies in the
     Philosophy of Science 37. Synthese Library 79.
     Dordrecht, Holland; Boston: D. Reidel Pub. Co.,
     1977.

> Contains's Helmholtz's essay "On the origin

and significance of the axioms of geometry"
(1870).

Henderson, Linda Dalrymple. *The Fourth Dimension and Non-Euclidean Geometry in Modern Art*. Princeton: Princeton University Press, 1983.

Hilbert, David. *Foundations of Geometry*. 2d ed. Trans. Leo Unger. La Salle, Ill.: Open Court, 1971.

Translated from the tenth German edition, rev. and enl. by Dr. Paul Bernays.

Originally published as *Grundlagen der Geometrie* (Leipzig, 1899).

Hilbert, David. *Natur und Mathematisches Erkennen. Vorlesungen, gehalten 1919-1920 in Göttingen*. Ed. David E. Rowe. Boston; Basel: Birkhäuser Verlag, 1992.

Hoüel, Guillaume Jules. "Essai d'une exposition rationelle des principes fondamentaux de la géométrie élémentaire," *Archiv der Mathematik und Physik* 40 (1863): 171-211.

Kant, Immanuel. *Critique of Pure Reason*. Trans. Norman Kemp Smith. London: Macmillan, 1929.

Translation of *Kritik der reinen Vernunft* (Riga, 1781).

Kant, Immanuel. *Philosophical Correspondence, 1759–1799*. Ed. Arnulf Zweig. Chicago: University of Chicago Press, 1967.

Klein, Christian Felix. *Gesammelte Mathematische Abhandlungen* I. Berlin: Julius Springer, 1921-23.

Contains Klein's "Vergleichende Betrachtungen über neuere geometrische Forschungen,

the Erlangen Program, Deichert, Erlangen,"
(1872), pp. 460-497

Klügel, Abraham Gotthelf. *Conatuum praecipuorum theoriam parallelarum demonstrandi recensio. . . .* Göttingen: ex officina Schultziana curante F. A. Rosenbusch, 1763.

Legendre, Adrien Marie. *Éléments de géométrie, avec des notes.* Paris: chez Firmin Didot, 1794.

Lobachevskii, Nikolai Ivanovich. *Geometrical Researches on the Theory of Parallels.* Trans. George Bruce Halsted. Austin: The University of Texas, 1891.

Lobachevskii, Nikolai Ivanovich. *Geometrische Untersuchungen zur Theorie der Parallellinien.* Berlin: G. Fincke, 1840.

Lobachevskii, Nikolai Ivanovich. *Geometrische Untersuchungen zur Theorie der Parallellinien.* 2nd ed. Berlin: Mayer & Müller, 1887.

Lobachevskii, Nikolai Ivanovich. *Zwei geometrische Abhandlungen aus dem russischen uebersetzt, mit anmerkungen und mit einer biographie des verfassers.* Trans. Friedrich Engel. Leipzig: B. G. Teubner, 1898-99.

Lobachevskii, Nikolai Ivanovich. "Études géométriques des parallèles par J. N. Lobatchewsky, traduit de l'Allemand par J. Hoüel," *Mémoires de la Société des Sciences physiques et naturelles de Bordeaux* 4 (1866): 83-128.

Reprinted as: *Études géométriques sur la théorie des parallèles, . . . traduit de l'allemand*

par J. Hoüel . . . suivi d'un extrait de la corre-
spondance de Gauss et de Schumacher. Paris:
Gauthier-Villars, 1866.

Moulton, R. F. "A Simple non-Desarguesian Plane
Geometry," *Transactions of the American Math-*
ematical Society 3 (1902): 192–195

Newton, Sir Isaac. *The Principia: Mathematical Prin-*
ciples of Natural Philosophy . . . A New Transla-
tion. Trans. I. Bernard Cohen and Anne Whit-
man. Ed. I. Bernard Cohen. Berkeley : University
of California Press, 1999.

Pasch, Moritz. *Vorlesungen über neuere Geometrie.*
Leipzig: B. G. Teubner, 1882.

Peano, Giuseppe. "Sui fondamenti della geometria,"
Rivista di matematiche 4 (1894): 73.

Pieri, Mario. "Sui principi che reggiono la geometria di posizione," *Atti della Reale Accademia*
delle scienze di Torino (30) 1895: 54–108.

Pieri, Mario. "I Principii della Geometria di Posizione,
composti in sistema logico deduttivo," *Memorie*
della Reale Accademia delle Scienze di Torino 2
48 (1899): 1–62.

Poincaré, Henri. *Science et méthode.* Paris: Flam-
marion, 1908.

    Contains Poincaré's essay "L'invention math-
ématique" on pp. 43–63.

Poincaré, Henri. "Les fondements de la géométrie,"
Bulletin de la Société Mathématique de France 26
(1902): 249–272.

Also published on pp. 92–113 of vol. 11 of *Oeuvres de Henri Poincaré*. Paris: Gauthier-Villars et Cie, 1916 ff.

Pont, Jean-Claude. *L'aventure des parallèles: histoire de la géométrie non euclidienne, précurseurs et attardés*. Berne; New York: Peter Lang, 1986.

Riemann, Bernhard. "Ueber die Hypothesen welche der Geometrie zu Grunde liegen," *Abhandlungen der Königlichen Gesellschaft der Wissenschaften zu Göttingen* 13 (1854): 1-20.

Also published on pp. 304–319 in Bernhard Riemann. *Gesammelte mathematische Werke, wissenschaftlicher Nachlass und Nachträge: collected papers*. Ed. Raghavan Narasimhan. Teubner-Archiv zur Mathematik. Supplement 1. Berlin; New York: Springer-Verlag; Leipzig: B. G. Teubner, 1990.

Russo, Lucio. "The Definitions of Fundamental Geometric Entities Contained in Book I of Euclid's *Elements*," *Archive for History of Exact Sciences* 52.3 (1998): 195-219.

Sartorius von Waltershausen, Wolfgang. *Gauss: zum gedächtnis*. Leipzig: S. Hirzel, 1856. Repr. Wiesbaden: Martin Sandig, 1965.

Stäckel, Paul. *Wolfgang und Johann Bolyai geometrische Untersuchungen, mit Unterstützung der Ungarischen Akademie der Wissenschaften*. 2 vols. Urkunden zur geschichte der nichteuklidischen Geometrie 2. Leipzig; Berlin: B. G. Teubner, 1913.

Volume 1: Leben und Schriften der beiden Bolyai.

Volume 2: Stücke aus den Schriften der beiden Bolyai.

Toepell, Michael-Markus. *Über die Entstehung von David Hilberts "Grundlagen der Geometrie."* Studien zur Wissenschafts-, Sozial- und Bildungsgeschichte der Mathematik 2. Göttingen: Vandenhoeck & Ruprecht, 1986.

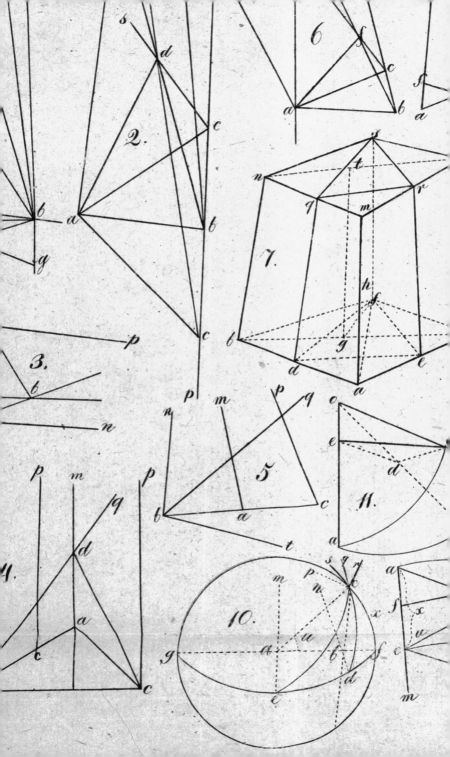

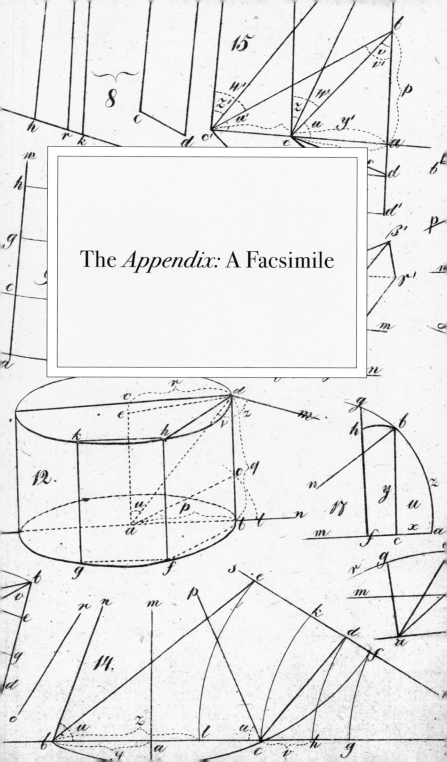

The *Appendix:* A Facsimile

APPENDIX.

SCIENTIAM SPATII *absolute veram* exhibens:

a veritate aut falsitate Axiomatis XI Euclidei
(a priori haud unquam decidenda) in-
dependentem; adjecta ad casum falsitatis, quadratura circuli
geometrica.

———— ·❦· ————

Auctore JOHANNE BOLYAI de eadem, Geometrarum
in Exercitu Caesareo Regio Austriaco Castrensium Capitaneo

————❦————

EXPLICATIO SIGNORUM.

$\widetilde{ab}$ denotet complexum *omnium* punctorum cum punctis a, b in recta sitorum.

$\widetilde{ab}$ - - - rectae $\widetilde{ab}$ in a bifariam sectae dimidium illud, quod punctum b complectitur.

$\widetilde{abc}$ - - - complexum *omnium* punctorum, quae cum punctis a, b, c (non in eadem recta sitis) in eodem plano sunt.

$ab\widetilde{c}$ - - - plani $\widetilde{abc}$ per $\widetilde{ab}$ bifariam secti dimidium, punctum c complectens.

abc - - - portionum, in quas $\widetilde{abc}$ per complexum rectarum $\widetilde{ba}$, $\widetilde{bc}$ dividitur, *minorem*; sive *angulum*, cuius $\widetilde{ba}$, $\widetilde{bc}$ crura sunt.

$abcd$ - - - (si d in abc sit et $\widetilde{ba}$, $\widetilde{cd}$ se invicem non secent) portionem ipsius abc inter $\widetilde{ba}$, $\widetilde{bc}$, $\widetilde{cd}$ comprehensam; $bacd$ vero portionem plani $\widetilde{abc}$ inter $\widetilde{ab}$, $\widetilde{cd}$ sitam.

L - - - perpendiculare.

$\wedge$ - - - angulum.

R - - - angulum rectum.

$ab \doteq cd$ - $cab = acd$.

$\equiv$ - - - congruens *).

$x \frown a$ - x tendere ad limitem a.

$\bigcirc r$ - - peripheriam circuli radii r.

$\odot r$ - - aream circuli radii r.

*) Sit fas, signo hocce, quo summus Geometra *GAVSS* *numeros congruos* insignivit; congruentiam geometricam quoque denotare: nulla ambiguitate exinde metuenda.

A

§ 1. (Fig.1.) Si rectam $\widetilde{am}$ non secet plani ejusdem recta $\widetilde{bn}$, at secet quaevis $\widetilde{bp}$ (in abn) : designetur hoc per $bn|||am$. *Dari* talem $\widetilde{bn}$, et quidem *unicam*, e quovis puncto b (extra $\widetilde{am}$), atque bam †abn non$>$ $2R$ esse patet; nam bc circa b mota, donec bam †$abc=2R$ fiat, $\widetilde{bc}$ ex $\widetilde{am}$ aliquando *primum* exit, estque tunc $bc|||am$. Nec non patet esse $bn|||em$, ubivis sit e in $\widetilde{am}$ (supponendo in omnibus talibus casibus esse $am > ae$). Et si, puncto c in $\widetilde{am}$ abeunte in infinitum, semper sit $cd = cb$: erit semper $cdb = (cbd < nbc)$; ast $nbc \frown o$; adeoque et adb $\frown o$.

§ 2. (Fig .2). Si $bn|||am$; est quoque $cn|||am$. Nam sit d ubicunque in $macn$. Si c in $\widetilde{bn}$ sit; $\widetilde{bd}$ secat $\widetilde{am}$ (propter $bn|||am$), adeoque et $\widetilde{cd}$ secat $\widetilde{am}$; si vero c in $\widetilde{bp}$ fuerit; sit $bq|||cd$: cadit $\widetilde{bq}$ in abn (§ 1), secatque $\widetilde{am}$, adeoque et $\widetilde{cd}$ secat $\widetilde{am}$. Quaevis $\widetilde{cd}$ igitur (in acn) secat in utroque casu $\widetilde{am}$ absque eo, ut $\widetilde{cn}$ ipsam $\widetilde{am}$ secet. Est ergo semper $cn|||am$.

§ 3. (Fig.2). Si tam br quam cs sit$|||am$, et c non sit in $\widetilde{br}$; tum $\widetilde{br}$, $\widetilde{cs}$ se invicem haud secant. Si enim $\widetilde{br}$, $\widetilde{cs}$ punctum d commune haberent; (per §. 2.) essent dr et ds simul$|||am$, caderetque (§.1.) $\widetilde{ds}$ in $\widetilde{dr}$ et c in $\widetilde{br}$ (contra hyp).

§ 4. (Fig.3). Si $man > mab$; pro quovis puncto b ipsius $\widetilde{ab}$ datur tale c in $\widetilde{am}$, ut sit $bcm = nam$. Nam datur per (§. 1.) $bdm > nam$, adeoque $mdp = man$, caditque b in $nadp$. Si igitur nam juxta am feratur, usquequo $\widetilde{an}$ in $\widetilde{dp}$ veniat; aliquando $\widetilde{an}$ per b transiisse, et aliquod $bcm = ncm$ esse oportet.

A 2

§ 5. (Fig. 1). Si $bn \mid\mid\mid am$, *datur* tale punctum f in $\widetilde{am}$, ut sit $fm \eqsim bn$. Nam per §. 1. datur $bcm >$ cbn; et si $ce = cb$, adeoque $ec \eqsim bc$; patet esse $bem < ebn$. Feratur p per ec, $\bigwedge lo$ bpm semper u, et $\bigwedge lo$ pbn semper v dicto; patet u esse prius ei simultaneo v minus, posterius vero esse majus. Crescit vero u a bem usque bcm *continuo*; cum (per §. 4.) *nullus* $\bigwedge lus > bem$ et $< bcm$ detur, cui u aliquando $=$ non fiat; pariter decrescit v ab ebn usque cbn continuo: datur itaque in ec tale f, ut $bfm = fbn$ sit.

§ 6. Si $bn \mid\mid\mid am$, atque ubivis sit e in $\widetilde{am}$, et g in $\widetilde{bn}$: tum $gn \mid\mid\mid em$ et $em \mid\mid\mid gn$. Nam (per §. 1.) est $bn \mid\mid\mid em$, et hinc (per §. 2.) $gn \mid\mid\mid em$. Si porro $fm \eqsim bn$ (§. 5.); tum $mfbn = nbfm$, adeoque (cum $bn \mid\mid\mid fm$ sit) etiam $fm \mid\mid\mid bn$, et (per praec.) $em \mid\mid\mid gn$.

§ 7. (Fig. 4). Si tam bn quam cp sit $\mid\mid\mid am$, et c non sit in $\widetilde{bn}$: est etiam $bn \mid\mid\mid cp$. Nam $\widetilde{bn}$, $\widetilde{cp}$ se invicem non secant (§. 3); sunt vero am, bn, cp aut in plano, aut non; atque in casu primo am aut in $bncp$ est, aut non. Si am, bn, cp in plano sint, et am in $bncp$ cadat: tum quaevis $\widetilde{bq}$ (in nbc) secat $\widetilde{am}$ in aliquo puncto d (quia $bn \mid\mid\mid am$); porro cum $dm \mid\mid\mid cp$ sit (§. 6.), patet $\widetilde{dq}$ secare $\widetilde{cp}$, adeoque esse $bn \mid\mid\mid cp$. Si vero bn, cp in eadem plaga ipsius am sint; tum aliqua earum ex. gr. cp, *intra* duas reliquas $\widetilde{bn}$, $\widetilde{am}$ cadit; quaevis $\widetilde{bq}$ (in nba) auem secat $\widetilde{am}$, adeoque et ipsam $\widetilde{cp}$. Est itaque $bn \mid\mid\mid cp$.

Si mab, mac, $\bigwedge lum$ efficiant; tum cbn cum abn nonnisi $\widetilde{bn}$, $\widetilde{am}$ vero (in abn) cum $\widetilde{bn}$, adeoque nbc quoque cum $\widetilde{am}$, nihil commune habent. Per quamvis $\widetilde{bd}$ (in nba) autem positum $bc\widetilde{d}$ secat $\widetilde{am}$, quia (propter $bn \mid\mid\mid am$) $b\widetilde{a}$ secat $\widetilde{am}$. Moto itaque bca cir-

ca bc, donec ipsam $\widetilde{am}$ *prima vice* deserat, postre-
mo cadet $\widetilde{bcd}$ in $\widetilde{bcn}$. Eadem ratione cadet idem
in $\widetilde{bcp}$; cadit igitur bn in bcp. Porro si $br \,|||\, cp$;
tum (quia etiam $am \,|||\, cp$) pari ratione cadit br in
bam; nec non (propter $br \,|||\, ep$) in bcp. Itaque $\widetilde{br}$
ipsis mab, pcb commune, nempe ipsum $\widetilde{bn}$ est, at-
que hinc $bn \,|||\, cp$.

Si igitur $cp \,|||\, am$, et b extra $\widetilde{cam}$ sit: tum sec-
tio ipsorum bam, bcp, nempe $\widetilde{bn}$ est $|||$ tam ad am,
quam ad cp.

§ 8. (Fig. 5). Si $bn \,|||\,$ et $\triangle\hspace{-0.5em}= cp$ (vel brevius $bn \,|||\, \triangle\hspace{-0.5em}=$
cp), atque ami (in $nbcp$) rectam $bc \,\llcorner\,$ riter bisse-
cet; tum $bn \,|||\, am$. Si enim $\widetilde{bn}$ secaret $\widetilde{am}$, etiam
$\widetilde{cp}$ secaret $\widetilde{am}$ in eodem puncto (cum $mabn = macp$),
quod et ipsis $\widetilde{bn}$, $\widetilde{cp}$ commune esset, quamvis bn
$|||\, cp$ sit. Quaevis $\widetilde{bq}$ (in cbn) vero secat $\widetilde{cp}$; adeo-
que secat $\widetilde{bq}$ etiam $\widetilde{am}$. Consequenter $bn \,|||\, am$.

§ 9. (Fig. 6). Si $bn \,|||\, am$, $map \,\llcorner\, mab$, atque $\wedge$,
quem nbd cum nba (in ea plaga ipsius $mabn$, ubi
map est) facit, sit $< R$: tum map et nbd se invicem secant. Nam sit $bam = R$, $ac \,\llcorner\, bn$ (sive in b
cadat c, sive non) et $ce \,\llcorner\, bn$ (in nbd); erit (per
hyp.) $ace < R$, et $af (\,\llcorner\, ce)$ in ace cadet. Sit $\widetilde{ap}$
sectio (punctum a commune habentium) $\widetilde{abf}$ et $\widetilde{amp}$;
erit $bap = bam = R$ (cum sit $bam \,\llcorner\, map$). Si denique $\widetilde{abf}$ in $\widetilde{abm}$ ponatur (a et b manentibus); ca-
det $\widetilde{ap}$ in $\widetilde{am}$; atque cum $ac \,\llcorner\, bn$ et $af \,\llcorner\, ac$ sit,
patet af *intra* $\widetilde{bn}$ terminari, adeoque bf in abn
cadere. Secat autem $\widetilde{bf}$ ipsam $\widetilde{ap}$ in *hoc* situ
quia $bn \,|||\, am$), adeoque etiam in situ *primo*, $\widetilde{ap}$ et
$\widetilde{f}$ se invicem secant; estque punctum sectionis ip-

sis $\widetilde{map}$ et $\widetilde{nbd}$ commune: secant itaque $\widetilde{map}$ et $\widetilde{nbd}$
se invicem. Facile exhinc sequitur $\widetilde{map}$ et $\widetilde{nbd}$ se
mutuo secare, si summa internorum, quos cum $mabn$
efficiunt, $< 2R$ sit.

§ 10. (Fig. 7). Si tam bn quam cp sit $|||\, \eqsim \, am$;
est etiam $bn|||\,\eqsim\, cp$. Nam mab et mac aut $\bigwedge lum$
efficiunt, aut in plano sunt.

Si prius; bissecet $\widetilde{qdf}$ rectam a^b $\llcorner riter$; erit
$dq\llcorner ab$, adeoque $dq \,|||\, am$ (§. 8.); pariter si $\widetilde{ers}$
bissecet rectam $ac \llcorner riter$, est $er |||\, am$; unde dq
$|||\, er$ (§. 7.). Facile hinc (per §. 9.) consequitur,
$\widetilde{qdf}$ et $\widetilde{ers}$ se mutuo secare, et sectionem $\widetilde{fs}$ esse
$|||\, dq$ (§. 7.), atque (propter $bn|||\, dq$) esse etiam
$fs|||\, bn$. Est porro (pro quovis puncto ipsius $\widetilde{fs}$)
$f^b = fa = fc$, caditque $\widetilde{fs}$ in planum $\widetilde{tgf}$, rectam bc
$\llcorner riter$ bissecans. Est vero (per §. 7.) (cum sit
$fs|||\, bn$) etiam $gt|||\, bn$. Pari modo demonstratur
$gt ||| \, cp$ esse. Interim gt bissecat rectam $bc\llcorner riter$;
adeoque $tbgn \eqsim tgcp$ (§. 1.) et $bn|||\, \eqsim\, cp$.

Si bn, am, cp in plano sint; sit (*extra* hoc pla-
num cadens) $fs|||\,\eqsim am$; tum (per praec.) $fs|||\,\eqsim$
tam ad bn quam ad cp, adeoque et $bn|||\,\eqsim cp$.

§ 11. Complexus puncti a, atque *omnium* puncto-
rum, quorum quodvis b tale est, ut si $bn|||\, am$
sit, sit etiam $bn \eqsim am$; dicatur F: sectio vero ipsius F cum quovis plano rectam am complectente
nominetur L. In quavis recta, quae $|||\, am$ est, F
gaudet puncto, et nonnisi uno; atque patet L per
am dividi in duas partes congruentes; dicatur $\widetilde{am}$
axis ipsius L; patet etiam, in quovis plano rectam am complectente, pro axe $\widetilde{am}$ unicum L dari.
Quodvis eiusmodi L, dicatur L ipsius $\widetilde{am}$ (in plano,
de quo agitur, intelligendo). Patet per L circa $\widetilde{am}$ revo-
lutum, F describi, cuius $\widetilde{am}$ axis vocetur, et vicissim
F axi $\widetilde{am}$ attribuatur.

§ 12. Si b ubivis in L ipsius $\widetilde{am}$ fuerit, et $bn|||\,\widetilde{am}$ ($\S$.11); tum L ipsius $\widetilde{am}$ et L ipsius $\widetilde{bn}$ coincidunt. Nam dicatur L ipsius $\widetilde{bn}$ distinctionis ergo l; sitque c ubivis in l, et $cp|||\,\widetilde{bn}$ ($\S$.11.); erit (cum et $bn|||\,\widetilde{am}$ sit) $cp|||\,\widetilde{am}$ ($\S$. 10), adeoque c etiam in L cadet. Et si c ubivis in L sit, et $cp|||\,\widetilde{am}$; tum $cp|||\,\widetilde{bn}$ ($\S$. 10.); cadit que c etiam in l ($\S$. 11). Itaque L et l sunt eadem; ac quaevis $\widetilde{bn}$ est etiam axis ipsius L, et inter omnes axes ipsius L, $\widetilde{=}$ est. Idem de F eodem modo patet.

$\S$. 13. (Fig.8). Si $bn|||\,am$, $cp|||\,dq$, et $bam\dagger abn$ $=2R$ sit; tum etiam $dcp\dagger cdq=2R$. Sit enim $ea=eb$ et $efm=dcp$ ($\S$.4.); erit (cum $bam\dagger abn=2R=abn$ $\dagger abg$ sit) $ebg=eaf$; adeoque si etiam $bg=af$ sit, $\triangle ebg\equiv\triangle eaf$, $beg=aef$, cadetque g in $\widetilde{fe}$. Est porro $gfm\dagger fgn=2R$ (quia $egb=efa$). Est etiam $gn|||\,fm$ ($\S$. 6.); itaque si $mfrs\equiv pcdq$, tum $rs|||\,gn$ ($\S$.7.), et r in vel extra fg cadit (si cd non $=fg$, ubi res jam patet).

I. In casu primo est frs non $>(2R-rfm=fgn)$, quia $rs|||\,fm$; ast cum $rs|||\,gn$ sit, est etiam frs non $<fgn$; adeoque $frs=fgn$, et $rfm\dagger frs=gfm\dagger$ $fgn=2R$. Itaque et $dcp\dagger cdq=2R$.

II. Si r extra fg cadat; tunc $ngr=mfr$, sitque $mfgn\equiv nghl\equiv lhko$ et ita porro, usquequo $fk=$ vel prima vice $>fr$ fiat. Est heic $ko|||\,hl|||\,fm$ ($\S$.7.). Si k in r cadat; tum ko in rs cadit ($\S$.1.); adeoque $rfm\dagger frs=kfm\dagger fko=kfm\dagger fgn=2R$; si vero r in hk cadat, tum (per I.) est $rhl\dagger krs=2R=rfm\dagger frs$ $=dcp\dagger cdq$.

$\S$14. Si $bn|||\,am$, $cp|||\,dq$, et $bam\dagger abn<2R$ sit; tum etiam $dcp\dagger cdq<2R$. Si enim $dcp\dagger cdq$ non esset $<$, adeoque (per $\S$. 1.) esset $=2R$; tum (per $\S$.13.) etiam $bam\dagger abn=2R$ esset (contra hyp).

$\S$15. Perpensis $\S\S$.13. et 14. *Systema Geometriae, hypothesi veritatis Axiomatis Euclidei* XI. *insistens dicatur* Σ; *et hypothesi contrariae supers-*

structum sit S. *Omnia, quae expresse non dicen-*
tur, *in* Σ vel *in* S *esse* ; *absolute enuntiari*, i. e.
illa, *sive* Σ *sive* S *reipsa sit*, *vera asseri intel-*
ligatur.

§ 16. (Fig.5). Si am sit axis alicujus L ; tum L in
Σ *recta* $\llcorner am$ est. Nam sit e quovis puncto b ipsius L
axis bn ; erit in Σ $bam+abn = 2bam = 2R$, adeoque
$bam = R$. Et si c quodvis punctum in $\widetilde{ab}$ sit, at-
que $cp \,|||\, am$; est (per §. 13.) $cp \doteq am$, ¡adeoque c
in L (§. 11.)

In S vero *nulla* 3 puncta a, b, c ipsius L vel F
in recta sunt. Nam aliquis axium am, bn, cp (ex.gr.
am) intra duos reliquos cadit ; et tunc (per §. 14.)
tam bam quam $cam < R$.

§ 17. *L est etiam in* S *line a, et* F *super $\widetilde{fic}$ies.*
Nam (per §. 11.) quodvis planum ad axem am (per
punctum aliquod ipsius F) $\llcorner re$, secat ipsum F in
peripheria circuli, cuius planum (per §. 14.) ad
nullum alium axem $\widetilde{bn} \llcorner re$ est. Revolvatur F
circa bn ; manebit (per §. 12.) quodvis punctum ipsius F in F, et sectio ipsius F cum plano ad $\widetilde{bn}$
non $\llcorner ri$, describet superficiem : atqui F (per §.12),
quaecunque puncta a, b fuerint in eo, ita *sibi* con-
gruere poterit, ut a in b cadat ; est igitur F su-
perficies uniformis. Patet hinc (per §. 11. et 12)
L esse *lineam uniformem.*

§18. (Fig.7). *Cujusvis plani*, per punctum a ipsius
F ad axem am oblique positi, *sectio* cum F in S *peri-*
pheria circuli est. Nam sint a, b, c, 3 puncta hujus
sectionis, et bn, cp axes ; facient $ambn$, $amcp$
$\wedge lum$; nam secus planum (ex §. 16.) per a, b, c
determinatum ipsam am complecteretur (contra
hyp). Plana igitur, rectas ab, $ac \llcorner riter$ bissecan-
tia se mutuo secant (§. 10.) in aliquo axe $\widetilde{fs}$ (i-
psius F), atque $fb = fa = fc$. Sit $ah \llcorner fs$, et revolvatur fah circa fs ; describet a peripheriam radii ha,
per b et c euntem, et *simul* in F et $\widetilde{abc}$ sitam

nec Fet $a\tilde{b}c$ praeter $\odot$ *ha* quidquam commune habent (§. 16.). Patet etiam portione *fa* lineae L (tanquam radio) in F circa *f* mota ipsam $\odot$ *hk* describi.

§ 19. (Fig.5). $\llcorner ris$ *bt* ad axem *bn* ipsius L (in.
planum ipsius L cadens) est in S *tangens* ipsius L.
Nam L in $\tilde{bt}$ praeter *b* nullo puncto gaudet (§.14.),
si vero *bq* in *tbn* cadat, tum centrum sectionis pla-
ni per *bq* ad *tbn* $\llcorner ris$ cum F ipsius $\tilde{bn}$ (§.18.) manifesto in $\tilde{bq}$ locatur, et si *bc* diameter sit, patet $\tilde{bq}$
lineam L ipsius $\tilde{bn}$ in *c* secare.

§. 20. Per quaevis 2 puncta in F linea *L* determi-
natur (§. 11. et 18); atque (cum ex §§. 16. et 19 . L
$\llcorner$ ad omnes suos axes sit) quivis $\wedge$ L lineus in F,
$\wedge lo$ planorum ad F per *crura* $\llcorner rium$,$=$est.

§ 21. (Fig. 6). Duae lineae Lformes $\tilde{ap}$, $\tilde{bd}$ in,
eodem F, cum tertia Lformi *ab* summam inter-
norum $<$ 2R efficientes, se mutuo secant (per $\tilde{ap}$
in F intelligendo L per *a,p* ductum, per $\tilde{ap}$ vero dimidium illud eius ex *a* incipiens, in quod *p* cadit).
Nam si *am, bn* axes ipsius F sint;tum $\tilde{amp}$, $\tilde{bnd}$ secant
se invicem (§.9.); atque F secat eorundem sectio-
nem (per §§.7. et 11.); adeoque et $\tilde{ap}$, $\tilde{bd}$ se mutuo
secant.

Patet exhinc Axioma XI. et omnia, quae in Geometria Trigonometriaque (plana) asseruntur, *absolute* constare in F, rectarum vices lineis L subeunti-
bus: idcirco functiones trigonometricae abhinc eodem
sensu accipientur, quo inΣveniunt; et peripheria circuli, cuius radius Lformis$=r$ in F, est$=2\pi r$, et pa-
riter $\odot r$ (in F) $= \pi r^2$ (per π intelligendo $\frac{1}{2}$ $\odot 1$ in
F, sive notum 3,1415926 ---)

§ 22. (Fig.9. Si $\tilde{ab}$ fuerit L ipsius $\tilde{am}$, et *c* in
$\tilde{am}$; atque $\wedge$ *cab* (e recta $\tilde{am}$ et Lformi linea

$\widetilde{ab}$ compositus) feratur prius juxta $\widetilde{ab}$, tum juxta $\widetilde{ba}$ semper porro in infinitum: erit via $\widetilde{cd}$ ipsius c linea $\mathbf{L}$ ipsius cm. Nam (posteriori l dicta) sit punctum quodvis d in $\widetilde{cd}$, $dn \,|||\, cm$, et b punctum ipsius $\mathbf{L}$ in $\widetilde{dn}$ cadens; erit $bn \eqcirc am$, et $ac = bd$, adeoque $dn \eqcirc cm$, consequ. d in l. Si vero d in l et $dn \,|||\, cm$, atque b punctum ipsius $\mathbf{L}$ ipsi $\widetilde{dn}$ commune sit; erit $am \eqcirc bn$ et $cm \eqcirc dn$, unde manifesto $bd = ac$, cadetque d in viam puncti c, et sunt l et $\widetilde{cd}$ eadem. Designetur tale l per $l \,||\, \mathbf{L}$.

§. 23. (Fig.9) Si linea $\mathbf{L}$ formis $cdf \,||\, abe$ (§.22.), et $ab = be$, atque $\widetilde{am}, \widetilde{bn}, \widetilde{ep}$ sint axes; erit manifesto $cd = df$; et si quaelibet 3 puncta a, b, e fuerint ipsius $\widetilde{ab}$, ac $ab = n \cdot cd$. erit quoque $ae = n \cdot cf$; adeoque (manifesto etiam pro ab, ae, dc incommensurabilibus) $ab : cd = ae : cf$, estque $ab : cd$ ab ab *independens*, et per ac *prorsus determinatum*. Denotetur quotus iste, nempe $ab : cd$ litera majori eiusdem nominis (puta per $\mathbf{X}$), quo ac litera minuscula (ex.gr. x) insignitur.

24. Quaecunque x et y fuerint; est $X = Y^{\frac{v}{x}}$ (§.23) Nam aut erit alterum (ipsorum x, y) multiplum alterius (ex.gr. y ipsius x), aut non.

Si $y = nx$; sit $x = ac = cg = gh$ &, usque quo $ah = y$ fiat; sit porro $cd \,||\, gk \,||\, hl$; erit (§.23.) $X = ab : cd = cd : gk = gk : hl$; adeoque $\frac{ab}{hl} = \left(\frac{ab}{cd}\right)^{n}$, sive $Y =$

$X^n = X^{\frac{y}{x}}$. Si x, y multipla ipsius i sint, puta $x = mi$, et $y = ni$; est (per praec.) $X = I^m$, $Y = I^n$, consequ. $Y = X^{\frac{n}{m}} = X^{\frac{y}{x}}$. Idem ad casum incommensurabilitatis ipsorum x, y facile extenditur. Si vero fuerit $q = y - x$; erit manifesto $Q = Y : X$.

Nec non manifestum est, in Σ pro quovis x es-

se $X = 1$, in S vero $X \gt 1$ esse, atque pro *quibus-*
vis ab, abe dari tale $cdf \parallel abe$, ut sit $cdf = ab$, unde $ambn = amep$ erit, etsi hoc illius qualevis multiplum sit; quod singulare quidem est, sed absur-
ditatem ipsius S evidenter non probat.

§.25. (Fig. 10) *In quovis rectilineo* $\triangle lo$ *sunt pe-*
ripheriae radiorum lateribus aequalium, uti sinus
$\wedge$*lorum oppositorum.*

Sit enim $abc = R$, et $am \perp bac$, atque sint bn,
$cp \parallel \parallel am$; erit $cab \perp ambn$, adeoque (cum $cb \perp ba$
sit) $cb \perp ambn$, consequ. $cpbn \perp ambn$. Secet F
ipsius $\widetilde{cp}$, rectas $\widetilde{bn}$, $\widetilde{am}$ (respective) in $d, e,$ et
fascias $cpbn$, $cpam$, $bnam$ in lineis Lformibus
cd, ce, de; erit (§. 20.) $cde = \wedge lo$ ipsorum $ndc, nde,$
adeoque $= R$; atque pari ratione est $ced = cab$,
Est autem (per §.21.) in Llineo $\triangle ced$ (heic radio semper $= 1$ posito) $ec : dc = 1 : \sin dec = 1 : \sin$
cab. Est quoque (per §.21.) $ec : dc = \bigcirc ec : \bigcirc dc$
(in F) $= \bigcirc ac : \bigcirc bc$ (§. 18.); adeoque est etiam
$\bigcirc ac : \bigcirc bc = 1 : \sin cab$; unde assertum pro quovis $\triangle lo$ liquet.

§. 26. *In quovis sphaerico* $\triangle lo$ *sunt sinus la-*
terum, uti sinus $\wedge$*lorum iisdem oppositorum.*

Fig. 11. Nam sit $abc = R$, et $ced \perp$ad sphaerae
radium oa; erit $ced \perp aob$, et (cum etiam $boc \perp$
loa sit) $cd \perp ob$. In $\triangle\triangle ceo$, cdo vero est (per
§. 25.) $\bigcirc ec : \bigcirc oc : \bigcirc dc = \sin coe : 1 : \sin cod =$
$\sin ac : 1 : \sin bc$; interim (§.25.) etiam $\bigcirc ec : \bigcirc dc =$
$\sin cde : \sin ced$; Itaque $\sin ac : \sin bc = \sin cde$
$\sin ced$; est vero $cde = R = cba$, atque $ced = cab$.
Consequenter $\sin ac : \sin bc = 1 : \sin a$. E quo pro-
nanans *Trigonometria sphaerica*, *ab Axiomate*
XI *independenter stabilita est*

§. 27. (Fig. 12.) Si ac, bd sint $\perp ab$, et feratur
ab juxta $\widetilde{ab}$; erit (via puncti c dicta heic cd) : $ab =$
$\sin u : \sin v$. Nam sit $de \perp ca$; est in $\triangle\triangle ade,$
adb (per§25.) $\bigcirc ed : \bigcirc ad : \bigcirc ab = \sin u : 1 : \sin v$. Re
voluto $bacd$ circa ac, describetur $\bigcirc ab$ per b, $\bigcirc d$
er d; et via dictae cd denotetur heic per $\bigodot de$.

Sit porro polygonum quodvis bfg --- ipsi $\odot ab$ inscriptum; nascetur per plana ex omnibus lateribus bf, fg &, ad $\odot ab \llcorner ria$, in $\odot cd$ quoque figura polygonalis totidem laterum; et demonstrari ad instar §. 23 potest, esse $cd : ab = dh : bf = hk : fg$ &, adeoque $dh{+}hk$ & $: bf{+}fg$ & $= cd : ab$. Quovis laterum bf, fg --- ad limitem o tendente, manifesto $bf{+}fg$ --- $\frown$ $\bigcirc ab$, et $dh{+}hk$ --- $\frown$ $\bigcirc ed$. Itaque etiam $\bigcirc ed : \bigcirc ab = cd : ab$. Erat vero $\bigcirc ed : \bigcirc ab$ $= \sin u : \sin v$. Conseq. $cd : ab = \sin u : \sin v$.

Remoto ac a bd in infinitum, manet $cd : ab$ adeoque etiam $\sin u : \sin v$ *constans*; u vero $\frown$ R (§.1.), et si $dm {|||} bn$ sit, v $\frown$ z; unde fit $cd : ab$ $= 1 : \sin z$. Via dicta cd denotabitur per $cd \parallel ab$.

§. 28. (Fig. 13.) Si $bn {|||} \rightleftharpoons am$, et c in $\widetilde{am}$, atque $ac = x$ sit : erit X (§.23.) $= \sin u : \sin v$. Nam si cd et ae sint $\llcorner bn$, et $bf \llcorner am$; erit (ad instar §. 27.) $\bigcirc bf : \bigcirc cd = \sin u : \sin v$. Est autem evidenter $bf = ae$; quamobrem $\bigcirc ea : \bigcirc dc = \sin u : \sin v$. In superficiebus vero Fformibus ipsorum am et cm (ipsum $ambn$ in ab et cg secantibus) est (per §.21.) $\bigcirc ea : \bigcirc dc = ab : cg = X$. Est itaque etiam $X = \sin u : \sin v$.

§.29. (Fig.14.) Si $bam = R$, $ab = y$, et $bn {|||} am$ sit; erit in S, $Y = \cot \frac{1}{2} u$. Nam si fuerit $ab = ac$, et $cp {|||} am$ (adeoque $bn {|||} \rightleftharpoons cp$), atque $pcd = qcd$; datur (§.19.) $ds \llcorner \widetilde{cd}$, ut $ds {|||} cp$, adeoque (§. 1.) $dt {|||} cq$ sit. Si porro $be \llcorner \widetilde{ds}$; erit (§. 7.) $ds {|||} bn$, adeoque (§.6.) $bn {|||} es$, et (cum $dt {|||} cg$ sit) $bq {|||} et$; consequ. (§. 1.) $ebn = ebq$. Repraesententur, bcf ex L ipsius bn, et fg, dh, ck et el ex Lformibus lineis ipsorum ft, dt, cq et et; erit evidenter (§.22.) $hg = df = dk = he$; itaque $cg = 2ch = 2v$. Pariter patet, $bg = 2 bl = 2z$ esse. Est vero $bc = bg{-}cg$; quapropter $y = z{-}v$, adeoque (§.24.) $Y = Z : V$. Est demum (§. 28.) $Z = 1 : sin \frac{1}{2} u$, et $V = 1 : sin (R{-} \frac{1}{2} u)$ consequ. $Y = cot \frac{1}{2} u$.

§. 30. (Fig.15.) Verumtamen facile (ex §. 25) patet, resolutionem problematis *Trigonometriae planae* in S, peripheriae per radium expressae indigere; hoc vero rectificatione ipsius L obtineri potest. Sint ab, cm, $c'm' \perp \widetilde{ac}$, atque b ubivis in $\widetilde{ab}$; erit (§.25.) $\sin u : \sin v := \bigcirc p : \bigcirc y$, et $\sin u' : \sin v' = \bigcirc p : \bigcirc y'$; adeoque $\frac{\sin u}{\sin v} \cdot \bigcirc y = \frac{\sin u'}{\cos v'} \cdot \bigcirc y'$. Est vero (per §27) $\sin v : \sin v' = \cos u : \cos u'$; conseq. $\frac{\sin u}{\cos u} \bigcirc y = \frac{\sin u'}{\cos u'} \bigcirc y'$; seu $\bigcirc y : \bigcirc y' = $ tang $u' :$ tang $u = $ tang $w :$ tang w'. Sint porro cn, $c'n' \parallel ab$, et cd, $c'd'$ lineae Lformes ad $\widetilde{ab} \perp res$; erit (§.21.) etiam $\bigcirc y : \bigcirc y' = r : r'$, adeoque $r : r' = $ tang $w :$ tang w'. Crescat iam p ab a incipiendo in infinitum; tum $w \frown z$, et $w' \frown z'$; quapropter etiam $r : r' = $ tang $z :$ tang z'. *Constans* $r :$ tang z (ab r *independens*) dicatur i; dum $y \frown o$, est ($\frac{r}{y} = \frac{i \text{ tang } z}{y}$) $\frown 1$, adeoque $\frac{y}{\text{tang } z} \frown i$. Ex §. 29 fit tang $z = \frac{1}{2}$ $(Y - Y^{-1})$; itaque $\frac{2y}{Y - Y^{-1}} \frown i$, seu (§.24.) $\dfrac{2y I^{\frac{y}{i}}}{I^{\frac{2y}{i}} - 1} \frown i$.

Notum autem est, expressionis istius (dum $y \frown o$) limitem esse $\frac{i}{\log \text{ nat } I}$; est ergo $\frac{i}{\log \text{ nat } I} = i$, et $I = e = 2,7182818\cdots$, quae quantitas insignis hic quoque elucet. Si nempe abhinc i illam rectam denotet, cuius $I = e$ sit, erit $r = i$ tang z. Erat autem (§.21.) $\bigcirc y = 2\pi r$; est igitur $\bigcirc y = 2\pi i$ tang $z = \pi i (Y - Y^{-1}) = \pi i (e^{\frac{y}{i}} - e^{\frac{-y}{i}}) = \frac{\pi y}{\log \text{ nat } Y} (Y - Y^{-1})$ (per §.24.)

§. 31. (Fig.16.) Ad resolutionem omnium $\triangle$*lorum* rectangulorum rectilineorum trigonometricam (e qua omnium $\triangle$*lorum* resolutio in promtu est) in S,

3 aequationes sufficiunt : nempe (a, b cathetos , c hypotenusam , et α, β ∧ los cathetis oppositos denotantibus) aequatio relationem exprimens 1mo inter a, b, α; 2do inter a, α, β; 3tio inter a, b, c; nimirum ex his reliquae 3 per eliminationem prodeunt.

I. Ex §.25. et 30. est 1 : sin α $= (C - C^{-1}) : (A -$

$A^{-1}) = (e^{\frac{c}{i}} - e^{\frac{-c}{i}}) : (e^{\frac{a}{i}} - e^{\frac{-a}{i}})$ (aequatio pro α; c, a).

II. Ex §.27. sequitur (si β m ||| γ n sit) cos, α : sin β $=$ 1 : sin u; ex §. 29 autem fit 1 : sin $u = \frac{1}{2} (A +$ $A^{-1})$; itaque cos α : sin β $= \frac{1}{2} (A + A^{-1}) = \frac{1}{2}$ $(e^{\frac{a}{i}} + e^{\frac{-a}{i}})$ (aequatio pro α, β, a).

III. Si αα' ⌐ βαγ , atque ββ' et γγ' fuerint || αα', (§.27), atque β'α'γ' ⌐ αα'; erit manifesto (uti in (§. 27) $\frac{ββ'}{γγ'} = \frac{1}{\sin u} = \frac{1}{2} (A + A^{-1}); \frac{γγ'}{αα'} = \frac{1}{2}$ $(B + B^{-1})$, ac $\frac{ββ'}{αα'} = \frac{1}{2} (C + C^{-1})$; consequ. $\frac{1}{2}(C +$ $C^{-1}) = \frac{1}{2} (A + A^{-1}) . \frac{1}{2} (B + B^{-1})$, sive $(e^{\frac{c}{i}} + e^{\frac{-c}{i}})$ $= \frac{1}{4} (e^{\frac{a}{i}} . + e^{\frac{-a}{i}}) (e^{\frac{b}{i}} + e^{\frac{-b}{i}}$ (aequatio pro a, b, c).

§.32. Si γαδ $= R$, et βδ ⌐ aδ sit ; erit ◯ c : ◯ a $=$ 1 : sin α, et ◯ c : ◯ (d $=$ βδ) $=$ 1 : cos α, adeoque (◯ x² pro quovis x factum ◯ x. ◯ x denotante) manifesto ◯a² + ◯d² $=$ ◯ c² . Est vero (per §.27. et II.) ◯ d $=$ ◯ b. $\frac{1}{2} (A + A^{-1})$, consequ. $(e^{\frac{c}{i}} - e^{\frac{-c}{i}})^2 =$ $\frac{1}{4} \left[e^{\frac{a}{i}} + e^{\frac{-a}{i}} \right] . \left[e^{\frac{b}{i}} - e^{\frac{-b}{i}} \right] + \left[e^{\frac{-a}{i}} - e^{\frac{-a}{i}} \right]^2$, alia aequatio pro a, b, c, (cuius membrum 2dum facile ad formam *symmetriceam* seu *invariabilem* reducitur.) Denique ex $\frac{\cos \alpha}{\sin \beta} = \frac{1}{2}(A + A^{-1})$, atque $\frac{\cos \beta}{\sin \alpha} = \frac{1}{2}(B + B^{-1})$, fit (per III.)$\cot \alpha \cdot \cot \beta = \frac{1}{2}(e^{\frac{c}{i}} + e^{\frac{-c}{i}})$ (aequatio pro a, β, c).

§. 32. Restat adhuc modum *problemata* in S resolvendi breviter ostendere, quo (per exempla magis obvia) peracto, demum quid theoria haecce praestet, candide dicetur.

I. (Fig. 17.) Sit $\widetilde{ab}$ linea in plano, et $y = f(x)$ aequatio eius (pro coordinatis ⌐*ribus*), et quodvis incrementum ipsius z dicatur dz, atque incrementa ipsorum x, y, et areae u, eidem dz respondentia, respective per dx, dy, du denotentur; sitque $bh \parallel cf$, et exprimatur (ex §. 31.) $\frac{bh}{dx}$ per y, ac quaeratur ipsius $\frac{dy}{dx}$ *limes* tendente dx ad limitem b, (quod ubi eiusmodi limes quaeritur, subintelligatur): innotescet exinde etiam limes ipsius $\frac{dy}{bh}$, adeoque tang hba; eritque (cum hbc manifesto nec > nec < adeoque $= R$ sit), *tangens* in b ipsius bg per y determinata.

II. Demonstrari potest, esse $\frac{dz^2}{dy^2 + bh^2} \frown 1$; Hinc *limes* ipsius $\frac{dz}{dx}$, et inde z integratione (per x expressum) reperitur. Et potest lineae cuiusvis *in concreto datae* aequatio in S inveniri, e. g. ipsius L. Si enim $\widetilde{am}$ axis ipsius L sit; tum quaevis $\widetilde{cb}$ ex $\widetilde{am}$ secat L (cum (per §.19.) quaevis recta ex a praeter $\widetilde{am}$ ipsum L secet); est vero (si bn axis sit), $X = 1 : \sin cbn$ (§. 28.), atque $Y = \cot \frac{1}{2}$ cbn,(§.29) unde fit $Y = X + \sqrt{(X^2 - 1)}$,seu $e^{\frac{y}{i}} = e^{\frac{x}{i}} +$

$\sqrt{(e^{\frac{2x}{i}}-1)}$ aequatio quaesita. Erit hinc $\frac{dy}{dx}$ ⌢

$X(X^2-1)^{\frac{-1}{2}}$; atqui $\frac{bh}{dx}=1:\sin cbn=X$; adeo

que $\frac{dy}{bh}$ ⌢ $(X^2-1)^{\frac{-1}{2}}$; $1+\frac{dy^2}{bh^2}$ ⌢ $X^2 (X$

$-1)^{-1}, \frac{dz^2}{bh^2}$ ⌢ $X^2(X^2-1)^{-1}$, atque $\frac{dz}{dz}$ ⌢

$X(X^2-1)^{-1}, \frac{dz^2}{bh^2}$ ⌢ $X^2 (X^2-1)^{-1}$, atque $\frac{dz}{dx}$

⌢ $X^2(X^2-1)^{\frac{-1}{2}}$; unde per integrationem

invenitur $z=i(X^2-1)^{\frac{1}{2}}=i \cot cbn$ (uti §. 30.).

III. Manifesto $\frac{du}{d}$ ⌢ $\frac{hfcbh}{dx}$, quod (nonni

si ab y dependens) iam primum per y exprimen

dum est; unde u integrando prodit.

Si (Fig. 12.) $ab=p$, $ac=q$, et $cd=r$, atque $cabdc$

$=s$ sit; poterit (uti in II.) ostendi, esse $\frac{ds}{dg}$ ⌢ r

quod $=\frac{1}{2}p (e^{\frac{q}{i}}+e^{\frac{-q}{i}})$ atque integrando $s=$

$\frac{1}{2}pi(e^{\frac{q}{i}}-e^{\frac{-q}{i}})$. Potest hoc absque integratione

quoque deduci. Aequatione e. g, circuli (ex §. 31

III), rectae (ex § 31, II), sectionis coni (per praec.

expressis; poterunt areae quoque his lineis clau

sae exprimi.

Palam est, superficiem t ad figuram planam p

(in distantia q) illam, esse ad p in ratione poten

tiarum 2darum linearum homologarum, sive uti

$\frac{1}{4}(e^{\frac{q}{i}}+e^{\frac{-q}{i}})$: 1. Porro computum soliditatis pa

ri modo tractatum, facile patet duas integrationes

requirere, (cum et differentiale ipsum hic nonni

si per integrationem determinetur); et ante o.

mnia solidum a p et t ac complexu omnium re-
tarum ad p ⌊rium fines ipsorum p, t connectenti-
um, clausum quaerendum esse. Reperitur solidum
istud (tam per integrationem quam sine ea) $= \frac{1}{8}$

$pi \left[e^{\frac{2q}{i}} - e^{\frac{-2q}{i}} \right] + \frac{1}{2} pq$. Superficies quoque cor-

porum in S determinari possunt, nec non *curva-
turae*, *evolutae*, *evolventesque* linearum qualium-
vis &. Quod curvaturam attinet; ea in S aut
ipsius L est, aut per radium circuli, aut *distantiam* curvae ad rectam ‖lae ab hac recta, determi-
natur; cum e praecedentibus facile ostendi possit,
praeter L, lineas circulares, ac rectae ‖las, nullas
in plano alias lineas uniformes dari.

IV. Pro circulo est (uti in III.) $\frac{d \overset{\frown}{x}}{dx}$ ⌢ $\bigcirc x$,

unde (per §. 29.) integrando fit $\odot x = \pi i^2 \left[e^{\frac{x}{i}} - 2 \right.$

$\left. + e^{\frac{-x}{i}} \right].$

V. Pro area $cabdc = u$ (Fig. 9.) (linea L formi ab
$= r$, huic ‖la $cd = y$, ac rectis ac, $bd = x$ clausa)

est $\frac{du}{dx}$ ⌢ y; atque (§. 24.) $y = r e^{\frac{-x}{i}}$; adeoque

integrando) $u = ri \left(1 - e^{\frac{-x}{i}} \right)$. Crescente x in in-

finitum, fiet in S, $e^{\frac{-x}{i}}$ ⌢ o, adeoque u ⌢ ri.
Per *quantitatem* ipsius $mabn$, in posterum li-
nes iste intelligetur. Simili modo invenitur, quod
si p sit figura in F; spatium a p et complexu a-
rium e terminis ipsius p ductorum clausum $= \frac{1}{2}$
i sit.

VI. Si angulus ad centrum segmenti (Fig. 10)
sphaerae sit $2u$, peripheria circuli maximi sit p, et
arcus fc $\left(\bigwedge \text{li } u \right) = x$; erit $1 : \sin u = p : \bigcirc bc$ (§. 25).

et hinc $\bigodot bc = p \sin u$. Interim est $x = \dfrac{pu}{2\pi}$, ac dx $= \dfrac{pdu}{2\pi}$. Est porro $\dfrac{dz}{dx} \frown \bigodot bc$, et hinc $\dfrac{dz}{du} \frown \dfrac{F^2}{2\pi} \sin u$, unde (integrando) $z = \dfrac{\sin v\, u}{2\pi} p^2$. Cogitetur F in quod p (per meditullium f segmenti transiens) cadit; planis $\widetilde{fem}$, $\widetilde{cem}$ per af, ac ad F ⌐^{riter} positis, ipsumque in feg, ce secantibus; et considerentur L formis cd (ex c ad feg ⌐^{ris}) nec non L formis cf; erit $cef = u$ (§. 20.), et (§. 21.) $\dfrac{fd}{p} = \dfrac{\sin v\, u}{2\pi}$, adeoque $z = fd.p$. Ast (§. 21.) $p = \pi$. fdg; itaque $z = \pi.\ fd.\ fdg$. Est autem (§. 21.) $fd.\ fdg = fc.fc$; consequ $z = \pi.fc.fc = \bigodot fc$ in F. Sit iam (Fig. 14.) $bj = cj = r$; erit $= $ (§. 29.) $2r = i\ (Y - Y^{-1})$, adeoque (§. 21.) $\bigodot 2r$ (in F) $= \pi i^2\ (Y - Y^{-1})^2$. Est quoque (IV) $\bigodot 2y = \pi i^2.\ (Y^2 - 2 + Y^{-2})$; igitur $\bigodot 2r$ (in F) $= \bigodot 2y$, adeoque et $z = \bigodot 2y$, sive superficies z segmenti sphaerici aequatur circulo, chorda fc tanquam radio descripto. Hinc tota sphaerae superficies $= \bigodot fg = fdg.p = \dfrac{p^2}{\pi}$, suntque superficies sphaerarum, uti 2dae potentiae peripheriarum earundem maximarum.

VII. Soliditas sphaerae radii x in S reperitur simili modo $= \frac{1}{2} \pi i3\ (X^2 - X^{-2}) - 2\pi i^2 x$; superficies per revolutionem lineae cd (Fig. 12.) circa ab orta $= \frac{1}{2} \pi ip\ (Q^2 - Q^{-2})$, et corpus per $cabdc$ descriptum $= \frac{1}{2} \pi i^2 p\ (Q2 + Q-2)$. Quomodo vero omnia a (IV.) hucusque tractata, etiam absque integratione perfici possint brevitatis studio supprimitur.

Demonstrari potest, omnis expressionis literam i continentis (adeoque hypothesi, quod detur i.

nnixae) *limitem*, *crescente i in infinitum*, *expri-*
nere quantitatem plane pro Σ (adeoque pro hypothesi *nullius i*), *siquidem non eveniant aequatio-*
nes identicae. Cave vero intelligas putari, *syste-*
ma ipsum variari posse (quod omnino *in se et per*
se determinatum est) sed tantum *hypothesin*, qiod
successive fieri potest, donec non ad absurdum
perducti fuerimus. *Posito* igitur, quod in *tali* expressione litera *i* pro casu, si *S* esset re ipsa, *il-*
lam quantitatem unicam designet, cuius $I = i$ sit;
i vero *revera* Σ fuerit, *limes dictus* loco expressionis accipi *cogitetur*: manifesto *omnes* expres-
siones ex *hypothesi realitatis* ipsius S oriundae
hoc sensu) *absolute valent*, etsi *prorsus ignotum*
it, *num* Σ *sit*, *aut non sit*.

Ita e. g. ex expressione in §. 29. obtenta facile
et quidem *tam* differentiationis auxilio, quam *abs-*
que eo) valor notus pro Σ prodit $\bigcirc x = 2\pi$; ex I.
§. 31.) rite tractato, sequitur 1 : sin $\alpha = c : a$; ex
I. vero $\frac{\cos \alpha}{\sin \beta} = 1$, adeoque $\alpha + \beta = R$; aequatio *pri-*
sa in III. fit indentica, adeoque *valet* pro Σ, quam-
vis nihil in eo *determinet*; ex *secunda* autem fluit
$^2 = a^2 + b^2$. *Aequationes notae fundamentales tri-*
onometriae planae in Σ. Porro inveniuntur (ex
. 32.) pro Σ area et corpus in IV, utrumque $= i$;
x IV. $\bigodot x = \pi c^2$; (ex VII) sphaera radii $x = \frac{4}{3}$
$x^3 \&$. Sunt quoque theoremata ad finem (VI)
nuntiata manifesto *inconditionate vera*.

§. 33. Superest adhuc quid theoria ista sibi ve-
t, (in §. 32 promissum) exponere.
I. Num Σ aut S aliquod *reipsa* sit, indecisum
manet.
II. Omnia ex hypothesi *falsitatis* Ax. XI. de-
ucta (semper *sensu* §. 32. intelligendo) *absolute*
valent, adeoque *hoc sensu nulli hypothesi inni-*
untur. Habetur idcirco *trigonometria plana a*
riori in qua *solum* systema *ipsum ignotum adeo-*

que solummodo *absolutae* magnitudines expresio-
num incognitae manent, per *unicum* vero casum
notum, manifesto totum systema figeretur. Trigonometria sphaerica autem in §. 26. absolute stabi-
litur (Habeturque Geometria, Geometriae planae
in Σ prorsus analoga in F).

III. Si *constaret*, Σ esse, nihil hoc respectu
amplius incognitum esset; si vero *constaret non es-
se* Σ, tunc (§. 31.) (e.g.)e lateribus x, y et $\bigwedge$lo rectilineo ab iis intercepto, in *concreto datis* ma-
nifesto in se et per se impossibile esset $\triangle$lum absolute resolvere (i, e.) a priori determinare $\bigwedge$los
ceteros et *rationem lateris tertii* ad duo data ;
nisi X, Y determinentur, ad quod *in concreto* habere aliquod *a* oporteret, cuius A notum esset ;
atque tum *i unitas naturalis longitudinum* esset,
(sicuti *e* est basis logarithmorum naturalium). Si
existentia hujus *i* constiterit; quomodo ad usum
saltem quam exactissime construi possit, ostende-
tur.

IV. Sensu in I et II exposito patet, omnia in
spatio methodo recentiorum Analytica (intra justos
fines valde laudanda) absolvi posse.

V. Denique lectoribus benevolis haud ingra-
tum futurum est; pro casu illo quodsi non Σ sed
S re ipsa esset, circulo aequale rectilineum construi.

§. 34. (Fig.12.) Ex d ducitur $dm \,|||\, an$ modo se-
quenti. Fiat ex d, $db \llcorner an$; erigatur e puncto quovis aliquo a rectae $\overline{ab}$, $ac \llcorner an$ (in dba), et demit-
tatur $de \llcorner ac$; erit $\frown ed : \frown ab = 1 : \sin z$ (§. 27) si-
quidem *fuerit* $dm \,|||\, bn$. Est vero $\sin z$ non > 1,
adeoque ab non $> de$. Descriptus igitur quadraus
radio ipsi de aequali, ex a in bac, gaudebit puncto aliquo b vel o cum $b\tilde{d}$ communi. Priori in casu manifesto $z = R$; in posteriori vero erit (§.2)
$(\frown ao = \frown ed) : \frown ab = 1 : \sin aob$, adeoque $z = aob$.
S itaque fiat $z = aob$; erit $dm \,|||\, bn$.

§. 35. (Fig.18.) Si fuerit δ reipsa; ducetur re‑
cta ad $\wedge$li acuti crus unum. $\llcorner$ris, quae ad alte‑
rum $\mathop{|||}$ sit, hoc modo. Sit $am \llcorner bc$, et accipiatur
$cb = cc$ tam parvum (per §.19.), ut si ducatur $bn \mathop{|||} am$
§. 34.), sit $abn > \wedge$lo dato. Ducatur porro $cp \mathop{|||} am$
§. 34.), fiantque nbq, pcd utrumque $= \wedge$lo dato;
et $\widetilde{bq}$, $\widetilde{cd}$ se mutuo secabunt. Secet enim $b\widetilde{q}$(quod
per constr. in nbc cadit) ipsam $\widetilde{cp}$ in e; erit (pro‑
pter $bn \eqcirc cp$) $ebc < ecb$, adeoque $ec < eb$. Sint ef
$= ec$, $efr = ecd$, et $fs \mathop{|||} ep$; cadet fs in bfr. Nam
cum $bn \mathop{|||} ep$, adeoque $bn \mathop{|||} ep$, atque $bn \mathop{|||} fs$ sit;
erit (§. 14.) $fbn + bfs < (2R = fbn + bfr)$; itaque
$fs < bfr$. Quamobrem $\widetilde{fr}$ secat $\widetilde{ep}$, adeoque $\widetilde{cd}$ quo‑
que ipsam $\widetilde{eq}$ in puncto aliquo d. Sit iam $dg = dc$,
atque $dgt = dcp = gbn$; erit (cum $cd \eqcirc gd$ sit) $bn \eqcirc$
$t \eqcirc cp$. Si fuerit lineae L formis ipsius bn, pun‑
tum in $\widetilde{bq}$ cadens k (§. 19.), et axis kl; erit $bn \eqcirc$
l, adeoque $bkl = bgt = dcp$; sed etiam $kl \eqcirc cp$: ca‑
dit ergo k manifesto in g, estque $gt \mathop{|||} bn$. Si ve‑
ro ho ipsum $bg \llcorner^{riter}$ bissecet; erit $ho \mathop{|||} bn$ con‑
tructum.

§.36. (Fig.10.) Si fuerint data recta $\widetilde{cp}$ et planum
$\widetilde{mab}$, atque fiat $cb \llcorner \widetilde{mab}$, bn (in $b\widetilde{cp}$) $\llcorner bc$, et $cq \mathop{|||}$
n (§. 34.); sectio ipsius $\widetilde{cp}$ (si haec in bcq cadat)
cum $\widetilde{bn}$ (in $\widetilde{cbn}$), adeoque cum $\widetilde{mab}$ reperitur. Et
si fuerint data duo plana $p\widetilde{cq}$, $m\widetilde{ab}$, et sit $cb \llcorner$
$\widetilde{tab}$, $cr \llcorner p\widetilde{cq}$, atque (in $b\widetilde{cr}$) $bn \llcorner bc$, $cs \llcorner cr$; cadent
n in $m\widetilde{ab}$, et cs in $p\widetilde{cq}$; et sectione ipsarum
$\widetilde{n}, \widetilde{cs}$ (si detur) reperta, erit $\llcorner$ris in pcq per ean‑
em ad cs ducta, manifesto sesctio ipsorum $m\widetilde{ab}$,
$\widetilde{cq}$.

§. 37. (Fig. 7.) In $a\widetilde{m} \mathop{|||} bn$ reperitur tale a, ut sit

$am \triangleq bn$: si (per §. 34.) construatur extra $n\widetilde{b}m$, gt $|||bn$, et fiant $bg \llcorner gt$, $gc = gb$, atque $cp |||gt$; ponaturque $tg\widetilde{d}$ ita, ut efficiat cum $tg\widetilde{b}$ $\bigwedge$lum illi aequalem, quem $pc\widetilde{a}$ cum $pc\widetilde{b}$ facit; atque quaeratur (per §. 36.) sectio $\widetilde{dq}$ ipsorum $tg\widetilde{d}$, $nb\widetilde{a}$; fiatque $ba \llcorner dq$. Erit enimvero ob $\bigwedge$lorum L lineorum in F ipsius bn exortorum similitudinem (§.21.) manifesto $db = da$, et $am \triangleq bn$.

Facile hinc patet (L lineis per *solos terminos* datis) reperiri posse etiam · *terminos* proportionis 4tum ac medium, atque omnes constructiones geometricas, quae in Σ in plano fiunt, hoc modo in F *absque* XI. *Axiomate* perfici posse. Ita e. g. $4R$ in quotvis partes aequales geometrice dividi potest, si sectionem istam in Σ perficere licet.

§. 38. (Fig.14.) Si construatur (per §. 37.) e. g. $nbq = \frac{1}{3}R$, et fiat (per §. 35) in S ad $b\widetilde{q} \llcorner$ria $am ||| bn$, atque determinetur (per §. 37.) $jm \triangleq bn$; erit, si $ja = x$ sit, (§. 28.) $X = 1 : \sin \frac{1}{3}R = 2$, atque x *geometrice* constructum. Et potest nbq ita computari, ut ja ab i quovis dato minus discrepet, cum nonnisi $\sin nbq = \frac{1}{e}$ esse debeat.

§. 39. (Fig 19.) Si fuerint (in plano) pq et st, $||$ rectae mn (. 27.), et ab, cd sint $\llcorner$res ad mn aequales; manifesto est $\triangle dec \equiv \triangle bea$, adeoque $\bigwedge$li (forsan mixtilinei) ecp, *eat* congruent, atque $ec = ea$. Si porro $cf = ag$, erit $\triangle acf \equiv \triangle cag$, et utrumque *quadrilateri fagc* dimidium est. Si *fagc*, *hagh* duo eiusmodi quadrilatera fuerint ad *ag*, inter pq et st; aequalitas eorum (uti apud *Euclidem*), nec non $\bigwedge$lorum *agc*, *agh* eidem *ag* insistentium, verticesque in $\widetilde{pq}$ habentium aequalitas patet. Est porro $acf = cig$, $gcq = cga$, atque $acf + acg + gcq = 2R$

§. 32.), adeoque etiam $cag + acg + cga = 2R$: ita-
ue in quovis eiusmodi $\triangle$lo acg summa 3 $\bigwedge$lorum
$= 2R$. Sive in ag (quae $\parallel mn$) ceciderit autem re-
a ag, sive non; $\triangle$lorum rectilineorum agc, agh,
m ipsorum, quam summarum $\bigwedge$lorum ipsorun-
m, aequalitas in aperto est.

§. 40. (Fig. 20.) *Aequalia* $\triangle$la abc, abd (*abhinc
ctilinea*) *uno latere aequali gaudentia*, *summas*
lorum *aequales habent*. Nam dividat mn bifariam
m ac quam bc, et sit pq (per c) $\parallel mn$; cadet d
$\widetilde{pq}$. Nam si $b\widetilde{d}$ ipsum $\widetilde{mn}$ in 'puncto e, adeo-
ue (§. 39.) ipsum $\widetilde{pq}$ ad distantiam $ef = eb$ secet;
it $\triangle abc = \triangle abf$, adeoque et $\triangle abd = \triangle abf$, unde
in f cadit: si vero $b\widetilde{d}$ ipsum $\widetilde{mn}$ non secuerit,
c punctum, ubi $\llcorner$ris rectam ab bissecans ipsum
secat, atque $gs = ht$ ita, ut $\widetilde{st}$ productam $b\widetilde{d}$
puncto aliquo k secet (quod fieri posse modo
nili patet, ut §. 4.); sint porro $sl = sa$, $lo \parallel st$,
que o sectio ipsorum $b\widetilde{k}$ et $\widetilde{lo}$; esset tum $\triangle abl =$
abo (§.39.), adeoque $\triangle abc > \triangle abd$ (contra hyp).

§. 41. (Fig. 21.) *Aequalia* $\triangle\triangle abc$, def, *aequali-
s* $\bigwedge$lorum *summis gaudent*. Nam secet mn tam
, quam bc, ita pq tam df quam fe bifariam, et
, $rs \parallel mn$, atque $to \parallel pq$; erit $\llcorner$ris ag ad rs aut $=$
ri dh ad to, aut altera e. g. dh erit maior: in quo-
s casu $\bigcirc df$ e centro a cum $\widetilde{gs}$ punctum aliquod
commune habet, eritque (§. 39.) $\triangle abk = \triangle abc =$
def. Est vero $\triangle akb$ (per §. 40.) $\triangle$lo dfe, ac (per
39.) $\triangle$lo abc aequiangulum. Sunt igitur etiam
$\triangle abc$, def aequiangula.
In S *converti* quoque theorema potest. Sint enim
$\triangle abc$, def reciproce aequiangula, atque $\triangle bal =$
def; erit (per praec.) alterum alteri, adeoque
am $\triangle abc$ $\triangle$lo abl aequiangulum, et hinc mani-
sto $bcl + blc + cbl = 2R$. Atqui (ex §. 31.) cuiusvis

$\triangle$li $\wedge$lorum summa in S, est $< 2R$: cadit igitur in c.

§. 42. (Fig.22.) Si fuerit *complementum* summ$\ldots$ $\wedge$lorum $\triangle$li *abc* ad $2R$, u, $\triangle$li *def* vero v; est $\triangle a\ldots$: $\triangle def = u : v$. Nam si quodvis $\wedge$lorum *acg*, *ge*$\ldots$ *hcb*, *dfk*, *kfe* sit $= p$, atque $\triangle abc = mp$, $\triangle def\ldots$ np; sitque *s* summa $\wedge$lorum cuiusvis $\triangle$li quod$\ldots$ p est; erit manifesto $2R - u = ms - (m-1) 2R = 2\ldots$ $- m (2R - s)$, et $u = m (2R - s)$, et pariter $v = \ldots$ $(2R - s)$. Est igitur $\triangle abc : \triangle def = m : n = u : v$. $\ldots$ casum incommensurabilitatis $\wedge$lorum *abc*, *def* qu$\ldots$ que extendi facile patet.

Eodem modo demonstratur $\triangle$Ia in superfic$\ldots$ sphaerica esse uti *excessus* summarum $\wedge$lorum e$\ldots$ rundem supra $2R$. Si $2 \wedge$li $\triangle$li sphaerici recti f$\ldots$ erint, tertius z erit excessus dictus; est autem $\triangle\ldots$ istud (peripheria maxima p dicta) manifesto $= \frac{z}{2}\ldots$

$\frac{p^2}{2\pi}$ (§. 32. VI.); consequ quodvis $\triangle$, cuius $\wedge$lor$\ldots$

excessus $= z$, est $= \frac{zp^2}{4\pi^2}$.

§. 43. (Fig.15.) Jam *area* $\triangle$li rectilinei in S p$\ldots$ summam $\wedge$lorum exprimetur. Si *ab* crescat in inf$\ldots$ nitum : erit (§. 42) $\triangle abc : (R - u - v)$ constans. E$\ldots$ vero $\triangle abc \frown bacn$ (§. 32. V.), et $R - u - v \frown\ldots$ z (§. 1.); adeoque $bacn : z = \triangle abc : (R - u - v) = \ldots$ $bac'n' : z'$. Est porro manifesto $bdcn : bd'c'n' = r : \ldots$ $= \text{tang } z : \text{tang } z'$ (§. 30.). Pro $y' \frown o$ autem e$\ldots$ $\frac{bd'c'n'}{bac'n'} \frown 1$, nec non $\frac{\text{tang } z'}{z'} \frown 1$; consequ$\ldots$ $bdcn : bacn = \text{tang } z : z$. Erat vero (§. 32) $bdcn = \ldots$ $ri = i^2 \text{ tang } z$; est igitur $bacn = zi^2$. Quovis $\triangle$lo cu$\ldots$ ius $\wedge$lorum summae complementum ad $2R$, z est $\ldots$ in posterum breviter $\triangle$ dicto, erit idcirco $\triangle = z i\ldots$

Facile hinc liquet, quod si (Fig.14.) *or* ||| *ab*$\ldots$ et *ro* ||| *ab* fuerint; *area* inter $\widetilde{or}$, $\widetilde{si}$, $\widetilde{oe}$, compre$\ldots$

hensa (quae manifesto limes absolutus est areae triangulorum rectilineorum sine fine crescentium, seu ipsius $\triangle$ pro $z \frown 2R$), sit $= \pi i^2 = \odot i$, in F. Limite isto per $\square$ denotato, erit porro (Fig. 15) (per §.30) $\pi r^2 = \text{tang } z^2 \, \square = \odot r$ in F (§ 21) $= \odot s$ (per § 32. VI.), si chorda dc, s dicatur. Si jam radio dato s, circuli in plano (sive radio L formi circuli in F) $\llcorner$riter bisecto, construatur (per § 34) $db \; ||| \rightleftharpoons cn$; demissa $\llcorner$ri ca ad db, et erecta $\llcorner$ri cm ad ca; habebitur z; unde (per § 37) $\text{tang } z^2$, radio L formi ad lubitum pro unitate assumto, *geometrice determinari potest, per duas lineas uniformes ejusdem curvaturae* (quae solis terminis datis, constructis axibus, manifesto tanquam rectae commensurari, atque hoc respectu rectis aequivalentes spectari possunt).

Porro (Fig. 23) construitur quadrilaterum ex.gr. regulare $= \square$, ut sequitur. Sit $abc = R$, $bac = \frac{1}{2}R$, $acb = \frac{1}{4}R$, et $bc = x$; poterit X (ex § 31.II) per meras radices quadraticas exprimi, et (per §.37) construi: habitoque X, (per § 38, sive etiam 29 et 35) x ipsum determinari potest. Estque octulum $\triangle abc$ manifesto $= \square$, atque *per hoc, circulus planus radii s, per figuram rectilineam, et lineas uniformes ejusdem generis (rectis, quoad comparationem inter se, aequivalentes) geometrice quadratus; circulus F formis vero eodem modo complanatus: habeturque aut Axioma XI Euclidis verum, aut quadratura circuli geometrica;* si hucusque indecisum manserit, quodnam ex his duobus revera locum habeat. Quoties $\text{tang } z^2$ *vel* numerus integer *vel* fractio rationalis fuerit, (cujus ad simplicissimam formam reductae) denominator *aut* numerus primus formae $2^m + 1$ (cujus est iam $2 = 2^0 + 1$) *aut* productum fuerit e quotcunque primis hujus formae, quorum (ipsum 2, qui plus quotvis vicibus occurrere potest, excipiendo) quivis *semel* ut factor occurrit: per theoriam po-

lyponorum ill. *GAVSS* (praeclarum nostri in
omnis aevi inventum), etiam ipsi $\tang z^2 \square = \bigodot$
(et nonnisi pro talibus valoribus ipsius z) figuram
rectilineam aequalem constituere licet. Nam *div
sio* ipsius $\square$ (theoremate § 42 facile ad quaelib
polygoma extenso *)* manifesto *sectionem* ipsius 2
requirit, quam (ut ostendi potest) unice lsub dic
conditione geometrice perficere licet. In omnib
autem talibus casibus praecedentia facile ad scopu
perducent. Et potest quaevis figura rectilinea i
polygonum regulare *n* laterum geometrice eonvert
siquidem *n* sub formam *GAVSSianam* cadat.

Superesset denique, (ut res omni numero abso
vatur *)* , impossibilitatem , (absque suppositior
aliqua) decidendi, num Σ, aut aliquod (et quoc
nam) S sit, demonstrare: quod tamen occasior
magis idoneae reservatur.

ERRATA.

§. 1. *l.* 6. pro ex $a\widetilde{m}$ primum exit, lege, primo non secat $a\widetilde{m}$.

§. 4. linea 2 pro $\widetilde{ab}$ lege $a\widetilde{b}$; *l.* 3. lege (per §. 1.), ultima *l.* lege *nam*;

Pag. 4. pro 6 lege §. 6; *l.* ult. pro $b\widetilde{a}$ lege $b\widetilde{d}$. Pro bissecare, lege ubique bisecare.

Pag. 5. *l.* 5. a calce, lege $af < ac$; penultima et ult. *l.* lege $a\widetilde{p}$ et $b\widetilde{f}$.

§. 7. Casu 3tio *praemisso* duo priores, adinstar casus 2di §. 10. brevius ac elegantius simul absolvi possunt.

§. 10. a calce, *l.* 4. lege *tgbn.*

§. 11. *l.* 7. et in calce, lege $a\widetilde{m}$;

Pag. 9. *l.* 2, pro prortione, lege, extremitate portionis.

§. 17. Demonstrationem ad S restringere haud necesse est; quum facile ita proponatur, ut absolute (pro S et Σ) valeat.

§. 19. penultima *l.* et ult. pro *c* lege *q*.

§. 20. *l.* 2 post 19 claudatur, linea penult. lege, L lineus.

§. 21. *l.* 1. deleatur comma post: in; et *l.* penult. lege $\frac{1}{2} \bigcirc$ l.

§. 22. post Fig. 9. claudatur.

§. 23. *l.* 4. lege, $ab = n.cd.$

§. 24. *l.* 1. lege $Y = X^{\frac{\gamma}{x}}$

Pag. 11. in calce lege $\bigcirc cd$, *l.* penult. lege $\bigcirc ed$.

Pag. 13. *l.* 7. et 8 lege $\frac{\sin u'}{\sin v} . \bigcirc y'$

Pag. 14. *l.* 4. lege a, c, α; linea 7 lege, pro

III. *l.* 3. lege $\frac{\gamma\gamma'}{\alpha\alpha'}$, linea penult. post $e^{\frac{-b}{i}}$ cla

datur; §. 32. deleatur.

Pag. 15. ante §. 32. *l.* penult. duae priores qua titates parenthesibus inclusae quadrari deben et primus terminus 3tiae exponentem positivu habere. *l.* ult. lege α, β, c

§. 32. I. *l.* 3. a calce, pro *hba*, lege *hbg*.

Pag. 16. *l.* 3. lege $\frac{dy}{bh}$, linea 4 lege, atque $\frac{dz}{bh}$; l

nea 5 lege $X(X^2-1)^{\frac{-1}{2}}$, et dele quod inter du

commata est. III. *l.* 1. lege $\frac{du}{ax}$; *l.* 5. lege $\frac{ds}{dg}$

l. 4. a calce, quantitas inclusa quadretur.

Pa. 17. VI. *l.* 1. post segmenti, insere, z:

Pa. 18. linea 12. pro $=(\S\ 29)$. lege $(\S\ 30)$; line 6. ante VII. dele $z = \odot\ 2y$, sive; et VI

linea 5, lege $\frac{1}{4}\ \pi i^2 p(Q - Q^{-1})^2$.

Pag. 19. *l.* 10. lege $=e$; linea 16. lege §. 30; li nea 13. a calce, lege, III. et in calce, pro ipsum lege, verum.

Pag. 20. *l.* 15. lege *beri* pro *bere*; linea 3. a cal ce, lege (§. 25.); linea 2. lege $= aob$;

Pag. 21. *l.* 2. post *unum*, dele punctum; et *l.* 3. a calce pro sesctio, lege sectio.

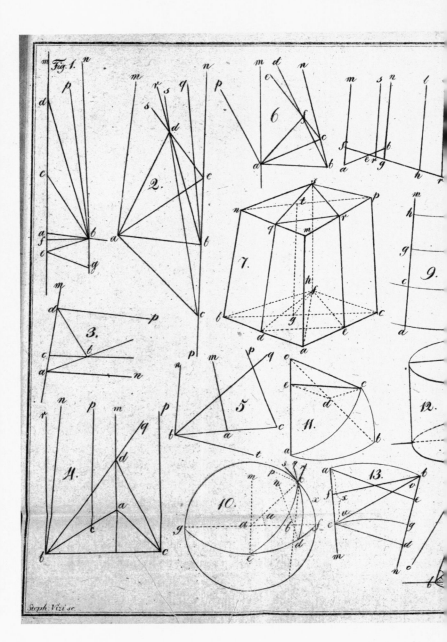

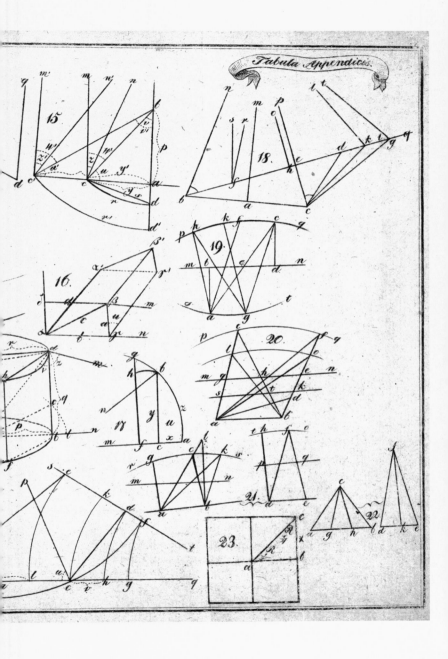

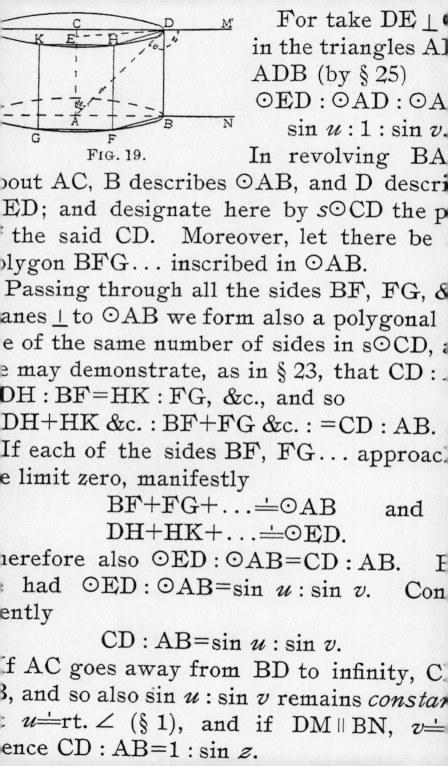

For take DE ⊥ (
in the triangles A[
ADB (by § 25)

⊙ED : ⊙AD : ⊙A[

sin u : 1 : sin v.

FIG. 19.　　In revolving BA[

[a]bout AC, B describes ⊙AB, and D descri[bes]
[⊙]ED; and designate here by s⊙CD the p[
[of] the said CD. Moreover, let there be [a]
[po]lygon BFG... inscribed in ⊙AB.

Passing through all the sides BF, FG, &[c.]
[pl]anes ⊥ to ⊙AB we form also a polygonal [
[lin]e of the same number of sides in s⊙CD, [
[w]e may demonstrate, as in § 23, that CD : [

[C]DH : BF = HK : FG, &c., and so

[C]DH + HK &c. : BF + FG &c. : = CD : AB.

If each of the sides BF, FG... approac[hes]
[th]e limit zero, manifestly

$$BF + FG + \ldots \doteq ⊙AB \qquad \text{and}$$
$$DH + HK + \ldots \doteq ⊙ED.$$

[Th]erefore also ⊙ED : ⊙AB = CD : AB.　　[
[we] had ⊙ED : ⊙AB = sin u : sin v.　　Con[sequ]-
[ent]ly

$$CD : AB = \sin u : \sin v.$$

[I]f AC goes away from BD to infinity, C[
[A]B, and so also sin u : sin v remains *constan[t]*
[:] $u \doteq$ rt. ∠ (§ 1), and if DM ∥ BN, $v \doteq$
[h]ence CD : AB = 1 : sin z.

The path called CD will be denoted b[y]
AB.

§ **28.** If BN ∥ ⨪ AM, and C in ray AM[,]
[A]C=x: we shall have (§ 23)

$$X=\sin u : \sin v.$$

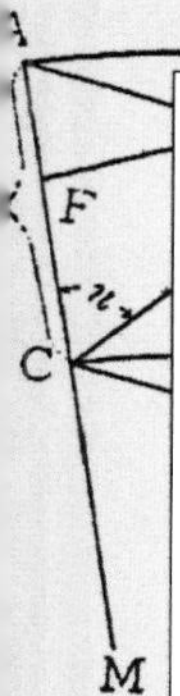

[Now if CD and A]C are ⊥ [...]
[...] [sh]all hav[e]

[...] u : sin v
[...] E: there[...]
[...] u : sin v
[...] surfac[e]
[A]M a[...] [A]B and
[b]y § [...]

$$\odot EA : \odot DC = AB : CG = X.$$

Therefore also

$$X = \sin u : \sin v.$$

§ **29.** If ∠BAM=rt.∠, and sect AB=y,

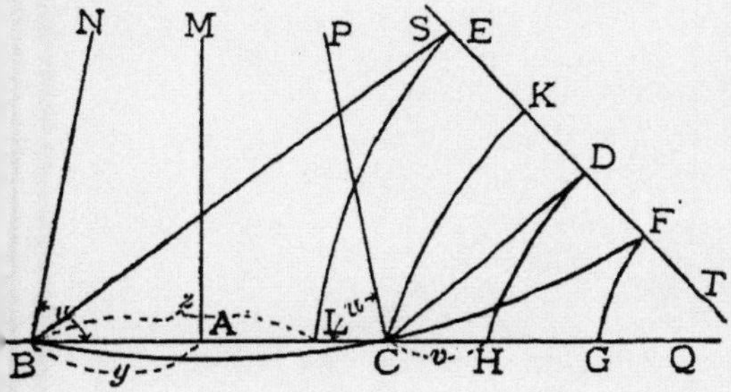

FIG. 21.

BN ∥ AM, we [...]
have in S

$$Y = \cotan \tfrac{1}{2} [...]$$

For, if sect A[...]
sect AC, and C[...]
AM (and so BN[...]
CP), and ∠PC[...]

QCD; there is given (§ 19) DS⊥ray CD[...]
[th]at DS ∥ CP, and so (§ 1) DT ∥ CQ. Moreo[ver]
BE ⊥ ray DS, then (§ 7) DS ∥ BN, and so

GEORGE BRUCE HALSTED

G EORGE BRUCE HALSTED, the translator of János Bolyai's essay on non-Euclidean geometry, will always be remembered as one of the most energetic popularizers of the new geometry. He also translated works by Nikolai Ivanovich Lobachevskii, Gerolamo Saccheri, and Henri Poincaré, and his translations have become the principal way into the subject for historically minded readers of English.

Born in Newark, New Jersey, on November 23, 1853, Halsted studied mathematics at Princeton University, which three previous generations of his family had also attended, graduating with an A.B. in 1875 and an A.M. in 1878. He then went on to The Johns Hopkins University, at the time the leading institution for mathematics in the United States, taking his Ph.D. in 1879 under the direction of the charismatic J. J. Sylvester. Halsted was Sylvester's first student, but he took the Ph.D. (for a dissertation entitled "Basis for a Dual Logic") *in absentia* because of a complicated situation that had arisen between Sylvester,

Halsted, and Peirce.[1] This episode was an early hint of Halsted's lifelong ability to get himself embroiled in problems; throughout his career he took strong positions and defended them in purple prose.

While at Johns Hopkins, Halsted published a lengthy bibliography on non-Euclidean geometry in the *American Journal of Mathematics*, the journal Sylvester had founded. And whatever the substance of the matter with Peirce might have been, Halsted's acquaintance with the well-connected English mathematician was his ticket to the world of European mathematics. He went to Berlin with a letter of introduction from Sylvester to Carl Borchardt, the editor of the leading mathematical journal, the *Journal für die reine und angewandte Mathematik*. There he stayed for a year before returning to Princeton, where he was first a tutor in mathematics, from 1879 to 1881, and then an instructor in postgraduate mathematics, from 1881 to 1884.[2] In 1881 Halsted published his first book, entitled *Mensuration, Metrical Geometry* (Boston, 1881), and in 1884 he moved to Austin and the University of Texas, where he stayed until 1903.

The Texas years were Halsted's most productive. He was not the first professor of mathematics at the University, but he stayed the longest of the early appointments, becoming the first professor of pure mathematics (on a salary of $3,500 initially) as the department grew and split into divisions for pure and applied mathematics. He was one of the fourteen founders of the Texas Academy of Sciences, which lasted from 1892 until 1914; his wife was one of the patrons, donating $500. He was also involved with the early years of *The American Mathematical Monthly*, the journal for college teachers of mathematics in the United States. The *Monthly* was the creation of Benjamin Franklin Finkel who, in the fall of 1893, began (with his colleague John M. Colaw) writing to high school teachers of mathematics and professors in the colleges and universities in order to solicit subscribers and contributions. Halsted was the first university professor to respond, sending a check for thirty dollars, an amount he contributed each year until he was fired from Texas. The first issue of the journal came out in January 1894, and Halsted not only sent money but contributed articles and solicited others. The first five volumes of the journal all have articles by him, mostly on non-Euclidean geometry. He was also heavily involved with Paul Carus's journal *The Monist*, and several of his translations were first published there as well as in the *Monthly*.

Halsted's dismissal from Texas was a consequence of his support for one of his best students, R. L. Moore, who went on to become a distinguished topologist and the inventor of the so-called Moore method of instruction. The Moore method dispenses with books in favor of a careful program of individual discovery in mathematics; naturally, it works better in some parts of the subject than others. Moore had come to the University of Texas in 1898, and took his M.A. in 1901. He stayed on at the university, where Halsted, inspired by Hilbert's *Grundlagen der Geometrie* (1899), was teaching a course on the axiomatic approach to geometry. Halsted set a problem that led to Moore showing that one of the axioms in Hilbert's account could be derived from the others, and so was not independent of them, as Hilbert had implied. Halsted arranged for Moore's discovery to be published in the *Monthly*, which brought it to the attention of the eminent E. H. Moore (no relation of R. L.) who was in charge of the mathematics department at the University of Chicago, by then the best in America. He had already come to the same conclusion, and independently included it in a published paper ("On the projective axioms of geometry," *Transactions of the American Mathematical Society* 3 (1902), 142–158), but he handsomely ceded priority to the lesser-known mathematician, and set about, with Halsted's support, bringing Moore to Chicago. Until there was a secure position for him to go to, however, Halsted sought to keep R. L. Moore at Texas, a plan that was blocked by the Regents of the University of Texas. After an exchange of letters Halsted went public with his praise for Moore (though he did not name him) and his objection to the Regents, who were political appointees. An article in *Science* in October 1902 and another in the *Educational Review* for December enraged the Regents, and they dismissed Halsted with a month's notice.

In 1904 Halsted, now a professor at Kenyon College, in Ohio, published his book *Rational Geometry*

Based on Hilbert's Foundations (New York, 1904), which is the first attempt to adapt Hilbert's axiomatic geometry to the teaching of geometry in an American university setting. Halsted had long been interested in rigor in geometry, and was often scathing about published lapses from this goal. The main point of his article in the *Educational Review* for December 1902 had been to argue for a rigorous, and for that reason, as he saw it, truly comprehensible account of his subject. Halsted agreed with Hilbert that greater rigor leads to greater simplicity and ease of understanding. He admitted that a first course in geometry does not call for full rigor and that generally a more intuitive account is to be preferred. Still, he said, existing texts do that, and there was need for at least one completely rigorous text. Halsted was criticized for this ambition by John Dewey, the educational theorist and the man with whom the philosophy of pragmatism is most often associated. Dewey had known Halsted from their student days at Johns Hopkins, and was now at Chicago. In a reply to Halsted in the *Educational Review* ("The Psychological and the Logical in Teaching Geometry") he observed that Halsted nowhere tried to help the reader pass from an imperfect, intuitive grasp of the subject to a rigorous one and so dispense with the prop of psychology, something Dewey thought impossible in any case. Dewey's criticism suggests that Halsted's approach can only work, if it works at all, for the most gifted aspiring mathematician.

Halsted continued working after his retirement, producing a translation of Saccheri's *Euclides Vindicatus* in 1920. He died in New York on March 16, 1922.

---

1. Parshall and Rowe. *The Emergence of the American Mathematical Research Community*, 132.
2. The date of 1894 given in the *Dictionary of Scientific Biography* article makes less sense than the 1884 date in the AMS account of the early years at the University of Texas.

Sources

Halsted, George Bruce. "Bibliography of Hyper-Space and Non-Euclidean Geometry," *American Journal of Mathematics* 1 (1878): 261-276; 384-385; and 2 (1879): 65-70.

Finkel, Benjamin Franklin. "The Human Aspect in the Early History of *The American Mathematical Monthly*, Address to the annual MAA meeting in Cleveland, 1930," *American Mathematical Monthly* 38 (1931): 305-320.

Lewis, A. C. "The Building of the University of Texas Mathematics Faculty, 1883-1938," in *A Century of Mathematics in America*. Ed. Peter Duren. III: 205-239. Providence: American Mathematical Society, 1989.

There is a picture of Halsted on p. 206.

Parshall, Karen Hunger and David Rowe. *The Emergence of the American Mathematical Research Community, 1876–1900: J. J. Sylvester, Felix Klein, and E. H. Moore*. History of mathematics 8. Providence: American Mathematical Society; London: London Mathematical Society, 1994.

THE SCIENCE ABSOLUTE OF SPACE

Independent of the Truth or Falsity of Euclid's Axiom XI (which can never be decided a priori).

BY

JOHN BOLYAI

TRANSLATED FROM THE LATIN

BY

DR. GEORGE BRUCE HALSTED
PRESIDENT OF THE TEXAS ACADEMY OF SCIENCE

FOURTH EDITION.

VOLUME THREE OF THE NEOMONIC SERIES

PUBLISHED AT
THE NEOMON
2407 Guadalupe Street
AUSTIN, TEXAS, U. S. A.
1896

APPENDIX.

SCIENTIAM SPATII *absolute veram* exhibens:

a veritate aut falsitate Axiomatis XI *Euclidei (a priori haud unquam decidenda) independentem:* adjecta ad casum falsitatis, quadratura circuli geometrica.

———•———

Auctore JOHANNE BOLYAI de eadem, Geometrarum in Exercitu Caesareo Regio Austriaco Castrensium Capitaneo.

EXPLANATION OF SIGNS.

The straight AB means the aggregate of all points situated in the same straight line with A and B.

The sect AB means that piece of the straight AB between the points A and B.

The ray AB means that half of the straight AB which commences at the point A and contains the point B.

The plane ABC means the aggregate of all points situated in the same plane as the three points (not in a straight) A, B, C.

The hemi-plane ABC means that half of the plane ABC which starts from the straight AB and contains the point C.

ABC means the smaller of the pieces into which the plane ABC is parted by the rays BA, BC, or the non-reflex angle of which the sides are the rays BA, BC.

ABCD (the point D being situated within ∠ ABC, and the straights BA, CD not intersecting) means the portion of ∠ ABC comprised between ray BA, sect BC, ray CD; while BACD designates the portion of the plane ABC comprised between the straights AB and CD.

⊥ is the sign of perpendicularity.

∥ is the sign of parallelism.

∠ means angle.

rt. ∠ is right angle.

st. ∠ is straight angle.

≅ is the sign of congruence, indicating that two magnitudes are superposable.

AB⪯CD means ∠ CAB= ∠ ACD.

$x_{=}a$ means x converges toward the limit a.

△ is triangle.

⊙r means the [circumference of the] circle of radius r.

area ⊙r means the area of the surface of the circle of radius r.

THE SCIENCE ABSOLUTE OF SPACE.

§1. If the ray AM is not cut by the ray [3] BN, situated in the same plane, but is cut by every ray BP comprised in the angle ABN, we will call ray BN *parallel* to ray AM; this is designated by BN ‖ AM.

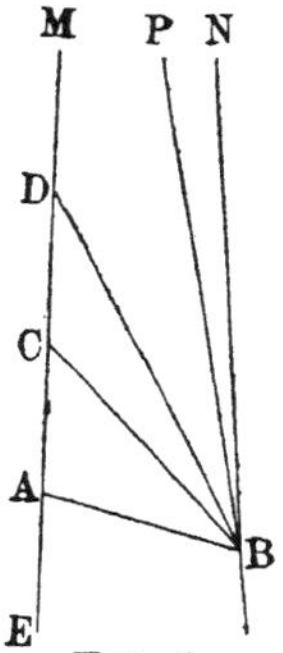

It is evident that *there is one such ray BN, and only one,* passing through any point B (taken outside of the straight AM), and that the sum of the angles BAM, ABN

Fig. 1.

can not exceed a st. ∠; for in moving BC around B until BAM+ABC=st. ∠, somewhere ray BC *first* does not cut ray AM, and it is then BC ‖ AM. It is clear that BN ‖ EM, wherever the point E be taken on the straight AM (supposing in all such cases AM>AE).

If while the point C goes away to infinity on ray AM, always CD=CB, we will have constantly CDB=(CBD<NBC); but NBC≐0; and so also ADB≐0.

§ 2. If BN ‖ AM, we will have also CN ‖ AM.
For take D anywhere in MACN. If C is on ray BN, ray BD cuts ray AM, since BN ‖ AM, and so also ray CD cuts ray AM. But if C is on ray BP, take BQ ‖ CD; BQ falls within the ∠ ABN (§1), and cuts ray AM; and so also ray CD cuts ray AM. Therefore every ray CD (in ACN) cuts, in each case, the ray AM, without CN itself cutting ray AM. Therefore always CN ‖ AM.

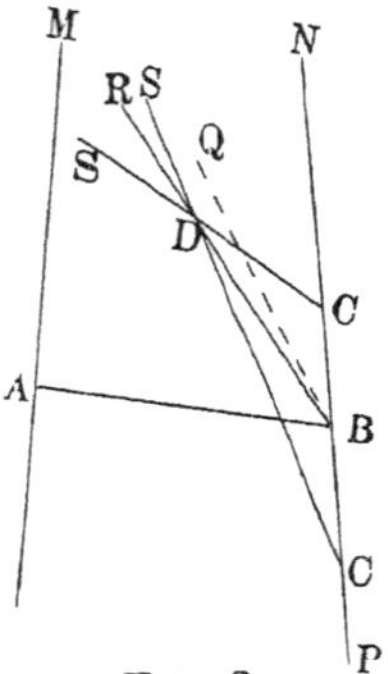

FIG. 2.

§ 3. (Fig. 2.) If BR and CS and each ‖ AM, and C is not on the ray BR, then ray BR and ray CS do not intersect. For if ray BR and ray CS had a common point D, then (§ 2) DR and DS would be each ‖ AM, and ray DS (§ 1) would fall on ray DR, and C on the ray BR (contrary to the hypothesis).

§ 4. If MAN>MAB, we will have for every point B of ray AB, a point C of ray AM, such that BCM=NAM.

For (by § 1) is granted BDM>NAM, and so that MDP=MAN, and B falls in

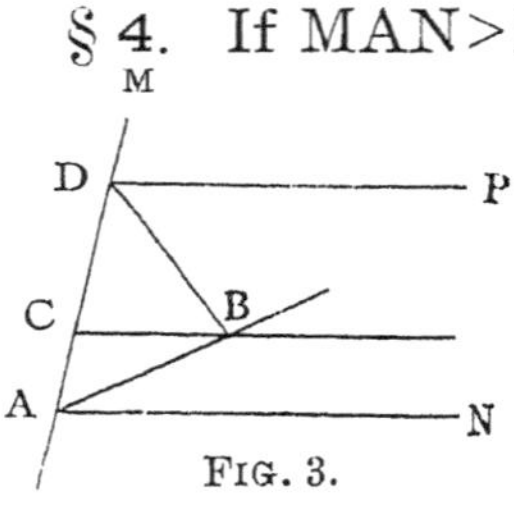

FIG. 3.

NADP. If therefore NAM is carried along AM until ray AN arrives on ray DP, ray AN will somewhere have necessarily passed through B, and some BCM=NAM.

§ 5. If BN ‖ AM, there is on the straight [4] AM a point F such that FM ≏ BN.

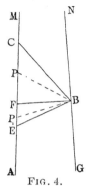

FIG. 4.

For by § 1 is granted BCM > CBN; and if CE=CB, and so EC ≏ BC; evidently BEM < EBN. The point P is moved on EC, the angle BPM always being called u, and the angle PBN always v; evidently u is at first less than the corresponding v, but afterwards greater. Indeed u increases *continuously* from BEM to BCM; since (by §4) there exists *no* angle >BEM and <BCM, to which u does not at some time become equal. Likewise v decreases continuously from EBN to CBN. There is therefore on EC a point F such that BFM=FBN.

§6. If BN ‖ AM and E anywhere in the straight AM, and G in the straight BN; then GN ‖ EM and EM ‖ GN. For (by §1) BN ‖ EM, whence (by §2) GN ‖ EM. If moreover FM ≏ BN (§5); then MFBN≅NBFM, and consequently (since BN ‖ FM) also FM ‖ BN, and (by what precedes) EM ‖ GN.

§ 7. If BN and CP are each ‖ AM, and C not on the straight BN; also BN ‖ CP. For the rays BN and CP do not intersect (§3); but AM, BN and CP either are or are not in the same plane; and in the first case, AM either is or is not within BNCP.

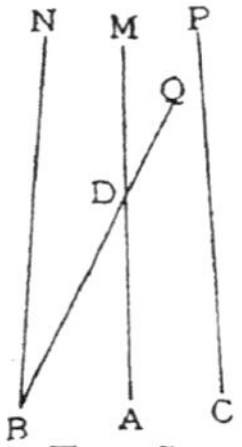

FIG. 5.   If AM, BN, CP are complanar, and AM falls within BNCP; then every ray BQ (in NBC) cuts the ray AM in some point D (since BN ‖ AM); moreover, since DM ‖ CP (§ 6), the ray DQ will cut the ray CP, and so BN ‖ CP.

But if BN and CP are on the same side of AM; then one of them, for example CP, falls between the two other straights BN, AM: but every ray BQ (in NBA) cuts the ray AM, and so also the straight CP. Therefore BN ‖ CP.

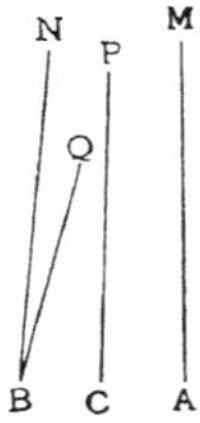

FIG. 6.   If the planes MAB, MAC make *an angle;* then CBN and ABN have in common nothing but the ray BN, while the ray AM (in ABN) and the ray BN, and so also NBC and the ray AM have nothing in common.

But hemi-plane BCD, drawn through any ray BD (in NBA), cuts the ray AM, since ray

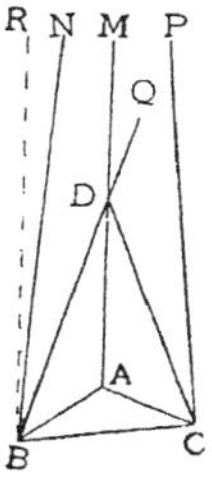

BQ cuts ray AM (as BN ‖ AM). Therefore in revolving the hemi-plane [5] BCD around BC until it *begins* to leave the ray AM, the hemi-plane BCD at last will fall upon the hemi-plane BCN. For the same reason this same will fall upon hemi-plane BCP.

FIG. 7. Therefore BN falls in BCP. Moreover, if BR ‖ CP; then (because also AM ‖ CP) by like reasoning, BR falls in BAM, and also (since BR ‖ CP) in BCP. Therefore the straight BR, being common to the two planes MAB, PCB, of course is the straight BN, and hence BN ‖ CP.*

If therefore CP ‖ AM, and B exterior to the plane CAM; then the intersection BN of the planes BAM, BCP is ‖ as well to AM as to CP.

§ 8. If BN ‖ and ≏ CP (or more briefly BN ‖ ≏ CP), and AM (in NBCP) bisects ⊥ the sect BC; then BN ‖ AM.

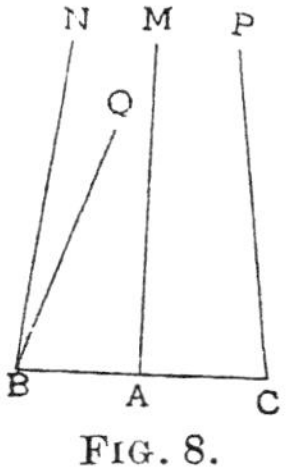

For if ray BN cut ray AM, also ray CP would cut ray AM at the same point (because MABN ≅ MACP), and this would be common to the rays BN, CP themselves, al-

FIG. 8.

---

* The third case being put before the other two, these can be demonstrated together with more brevity and elegance, like case 2 of § 10. [Author's note.]

though BN ‖ CP.   But every ray BQ (in CBN) cuts ray CP; and so ray BQ cuts also ray AM. Consequently BN ‖ AN.

§ 9.   If BN ‖ AM, and MAP ⊥ MAB, and the ∠, which NBD makes with NBA (on that side of MABN, where MAP is) is <rt.∠; then MAP and NBD intersect.

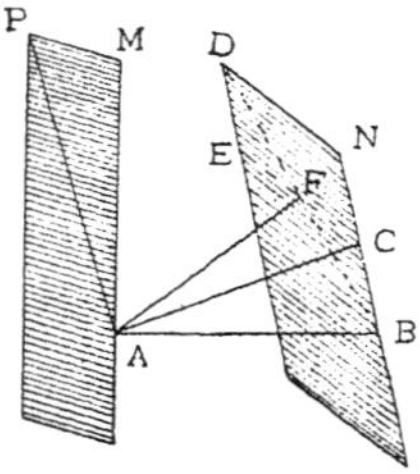

FIG. 9.

For let ∠BAM=rt.∠, and AC ⊥ BN (whether or not C falls on B), and CE ⊥ BN (in NBD); by hypothesis ∠ACE <rt.∠, and AF (⊥ CE) will fall in ACE.

Let ray AP be the intersection of the hemi-planes ABF, AMP (which have the point A common); since BAM ⊥ MAP, ∠BAP=∠BAM =rt.∠.

If finally the hemi-plane ABF is placed upon the hemi-plane ABM (A and B remaining), ray AP will fall on ray AM; and since AC ⊥ BN, and sect AF<sect AC, evidently sect AF will terminate within ray BN, and so BF falls in ABN. But in *this* position, ray BF cuts ray AP (because BN ‖ AM); and so ray AP and ray BF [6] intersect also in *the original* position; and the point of section is common to the hemi-planes MAP and NBD.   Therefore the hemi-planes MAP and NBD intersect.   Hence follows eas-

ily that the hemi-planes MAP and NBD intersect if the sum of the interior angles which they make with MABN is $<$ st.$\angle$.

§ 10. If both BN and CP $\parallel \eqsim$ AM; also is BN $\parallel \eqsim$ CP.

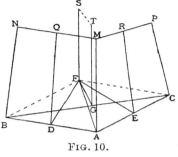

For either MAB and MAC make an angle, or they are in a plane.

If the first; let the hemi-plane QDF bisect $\perp$ sect AB; then DQ $\perp$ AB, and so DQ

FIG. 10.

$\parallel$ AM (§ 8); likewise if hemi-plane ERS bisects $\perp$ sect AC, is ER $\parallel$ AM; whence (§ 7) DQ $\parallel$ ER.

Hence follows easily (by § 9), the hemi-planes QDF and ERS intersect, and have (§ 7) their intersection FS $\parallel$ DQ, and (on account of BN $\parallel$ DQ) also FS $\parallel$ BN. Moreover (for any point of FS) FB=FA=FC, and the straight FS falls in the plane TGF, bisecting $\perp$ sect BC. But (by § 7) (since FS $\parallel$ BN) also GT $\parallel$ BN. In the same way is proved GT $\parallel$ CP. Meanwhile GT bisects $\perp$ sect BC; and so TGBN$\cong$ TGCP (§ 1), and BN $\parallel \eqsim$ CP.

If BN, AM and CP are in a plane, let (falling without this plane) FS $\parallel \eqsim$ AM; then (from what precedes) FS ∥ ≏ both to BN and to CP, and so also BN ∥ ≏ CP.

§ 11. Consider the aggregate of the point A, and *all* points of which any one B is such, that if BN ∥ AM, also BN ≏ AM; call it F; but the intersection of F with any plane containing the sect AM call L.

F has a point, and one only, on any straight ∥ AM; and evidently L is divided by ray AM into two congruent parts.

Call the ray AM *the axis* of L. Evidently also, in any plane containing the sect AM, there is for the *axis* ray AM a single L. Call any L of this sort the L of this ray AM (in the plane considered, being understood). Evidently by revolving L around AM we describe the F of which ray AM is called the axis, and in turn F *may be ascribed to the axis ray AM.*

[7]   § 12. If B is anywhere on the L of ray AM, and BN ∥ ≏ AM (§ 11); then the L of ray AM and the L of ray BN *coincide.* For suppose, in distinction, L′ the L of ray BN. Let C be anywhere in L′, and CP ∥ ≏ BN (§ 11). Since BN ∥ ≏ AM, so CP ∥ ≏ AM (§ 10), and so C also will fall on L. And if C is anywhere on L, and CP ∥ ≏ AM; then CP ∥ ≏ BN (§ 10); and C also falls on L′ (§ 11). Thus L and L′ are the

same; and every ray BN is also axis of L, and between all axes of this L, is $\eqcirc$.

The same is evident in the same way of F.

§13. If BN ‖ AM, and CP ‖ DQ, and∠BAM +∠ABN=st. ∠; then also ∠DCP+∠CDQ= st. ∠.

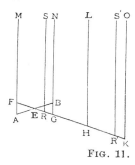

FIG. 11.

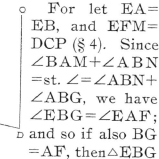

For let EA= EB, and EFM= DCP (§ 4). Since ∠BAM+∠ABN =st. ∠=∠ABN+ ∠ABG, we have ∠EBG=∠EAF; and so if also BG =AF, then△EBG ≅△EAF, ∠BEG=∠AEF and G will fall on the ray FE. Moreover ∠GFM+∠FGN=st. ∠ (since ∠EGB=∠EFA).

Also GN ‖ FM (§ 6).

Therefore if MFRS≅PCDQ, then RS ‖ GN (§ 7), and R falls within or without the sect FG (unless sect CD=sect FG, where the thing now is evident).

I. In the first case ∠FRS is not >(st. ∠−∠ RFM=∠FGN), since RS ‖ FM. But as RS ‖ GN, also ∠FRS is not <∠FGN; and so ∠FRS =∠FGN, and ∠RFM+∠FRS=∠GFM+∠

FGN=st.∠.   Therefore also ∠DCP+∠CDQ
=st.∠.

II. If R falls without the sect FG; then
∠NGR=∠MFR, and let MFGN≅NGHL≅
LHKO, and so on, until FK=FR or begins to
be >FR.   Then KO ‖ HL ‖ FM (§7).

If K falls on R, then KO falls on RS (§1);
and so ∠RFM+∠FRS=∠KFM+∠FKO=∠
KFM+∠FGN=st.∠; but if R falls within the
sect HK, then (by I) ∠RHL+∠KRS=st.∠=
∠RFM+∠FRS=∠DCP+∠CDQ.

§ 14.   If BN ‖ AM, and CP ‖ DQ, and ∠BAM
+∠ABN<st.∠; then also ∠DCP+∠CDQ<
st.∠.

For if ∠DCP+∠CDQ were not <st.∠, and
so (by §1) were =st.∠, then (by §13) also ∠
BAM+∠ABN=st.∠ (contra hyp.).

§15.   Weighing §§13 and 14, *the System of
Geometry resting on the hypothesis of the
truth of Euclid's Axiom XI is called Σ; and
the system founded on the contrary hypoth-*
[8] *esis is S.*

*All things which are not expressly said to
be in Σ or in S, it is understood are enunciated absolutely, that is are asserted true
whether Σ or S is reality.*

§ 16. If AM is the axis of any L; then L, in Σ is a straight $\perp$ AM.

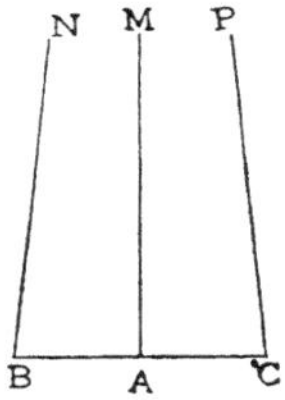

FIG. 12.

For suppose BN an axis from any point B of L; in Σ, $\angle$BAM+$\angle$ABN =st.$\angle$, and so $\angle$BAM=rt.$\angle$.

And if C is any point of the straight AB, and CP ‖ AM; then (by § 13) CP$\fallingdotseq$AM, and so C on L (§ 11).

But in S, no three points A, B, C on L or on F are in a straight. For some one of the axes AM, BN, CP (e. g. AM) falls between the two others; and then (by § 14) $\angle$BAM and $\angle$CAM are each <rt.$\angle$.

§ 17. *L in S also is a line, and F a surface.* For (by § 11) any plane $\bot$ to the axis ray AM (through any point of F) cuts F in [the circumference of] a circle, of which the plane (by § 14) is $\perp$ to no other axis ray BN. If we revolve F about BN, any point of F (by § 12) will remain on F, and the section of F with a plane not $\perp$ ray BN will describe a surface; and whatever be the points A, B taken on it, F can so be congruent to itse!f that A falls upon B (by § 12); therefore F is *a uniform surface.*

Hence evidently (by §§ 11 and 12) L is a uniform line.*

§ 18. *The intersection* with F of *any plane*, drawn through a point A of F obliquely to the axis AM, is, in S, *a circle*.

For take A, B, C, three points of this section, and BN, CP, axes; AMBN and AMCP make an angle, for otherwise the plane determined by A, B, C (from § 16) would contain AM, (contra hyp.). Therefore the planes bisecting ⊥ the sects AB, AC intersect (§ 10) in some *axis* ray FS (of F), and FB=FA=FC.

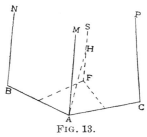

FIG. 13.

Make AH ⊥ FS, and revolve FAH about FS; A will describe a circle of radius HA, passing through B and C, and situated *both* in F and in [9] the plane ABC; nor have F and the plane ABC anything in common but ⊙ HA (§ 16).

It is also evident that in revolving the portion FA of the line L (as radius) in F around F, its extremity will describe ⊙ HA.

---

* It is not necessary to restrict the demonstration to the system S; since it may easily be so set forth. that it holds absolutely for S and for Σ.

§ 19. *The perpendicular* BT to the axis
BN of L (falling in the plane of L) is, in S,
tangent to L. For L has in ray
BT no point except B (§ 14),
but if BQ falls in TBN, then
the center of the section of the
plane through BQ perpendicular
to TBN with the F of ray BN
(§ 18) is evidently located on ray
BQ; and if sect BQ is a diameter, evidently
ray BQ cuts in Q the line L of ray BN.

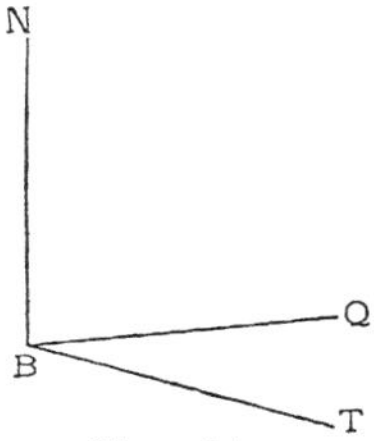

Fig. 14.

§ 20. Any two points of F determine a line
L (§§ 11 and 18); and since (from §§ 16 and 19)
L is ⊥ to all its axes, every ∠ of lines L in F is
equal to the ∠ of the planes drawn through its
sides perpendicular to F.

§ 21. Two L form lines, ray AP and ray
BD, in the same F, making with
a third L form AB, a sum of interior angles < st.∠, intersect.

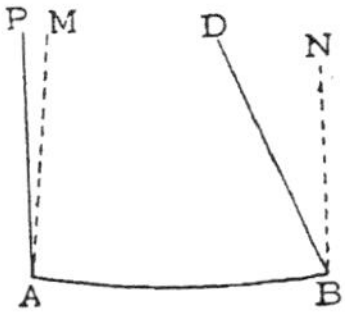

(By line AP in F, is to be
understood the line L drawn
through A and P, but by ray AP
that half of this line beginning at A, in which
P falls.)

Fig. 15.

For if AM, BN are axes of F, then the hemi-
planes AMP, BND intersect (§ 9); and F cuts

their intersection (by §§ 7 and 11); and so also ray AP and ray BD intersect.

From this it is evident that Euclid's Axiom *XI* and all things which are claimed in geometry and plane trigonometry hold good *absolutely* in F, L lines being substituted in place of straights: therefore the trigonometric functions are taken here in the same sense as in Σ; and the circle of which the L form radius $= r$ in F, is $= 2\pi r;$ and likewise area of $\odot r$ (in F) $= \pi r^2$ (by π understanding $\tfrac{1}{2}\odot 1$ in F, or the known $3.1415926\ldots$)

§ 22. If ray AB were the L of ray AM, and C on ray AM; and the ∠CAB (formed by the straight ray AM and the L form line ray AB), carried first along [10] the ray AB, then along the ray BA, always forward to infinity: the path CD of C will be the line L of CM.

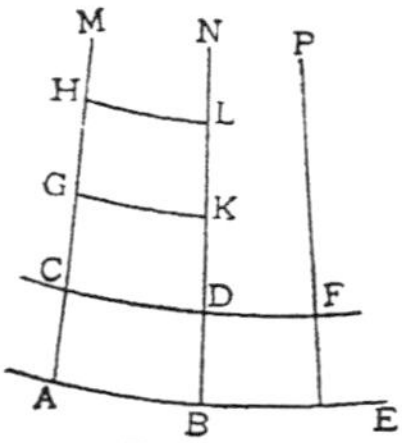

FIG. 16.

For let D be any point in line CD (called later L′), let DN be ‖ CM, and B the point of L falling on the straight DN. We shall have BN ≏ AM, and sect AC = sect BD, and so DN ≏ CM, consequently D in L′. But if D in L′ and DN ‖ CM, and B the point of L on the straight DN; we shall have AM ≏ BN and CM ≏ DN, whence manifestly sect BD = sect AC, and D will fall on the path of the point C, and L′ and the line CD are the same. Such an L′ is designated by L′∥L.

§ 23. If the L form line CDF ∥ ABE (§ 22), and AB=BE, and the rays AM, BN, EP are axes; manifestly CD=DF; and if any three points A, B, E are of line AB, and AB=n.CD, we shall also have AE=n.CF; and so (manifestly even for AB, AE, DC incommensurable), AB:CD=AE:CF, and AB:CD is *independent of AB, and completely determined by AC.*

This ratio AB:CD is designated by the capital letter (as X) corresponding to the small letter (as x) by which we represent the sect AC.

§ 24. Whatever be x and y; (§ 23), $Y = X^{\frac{y}{x}}$.

For, one of the quantities x, y is a multiple of the the other (e. g. y of x), or it is not.

If $y = n.x$, take $x = AC = CG = GH = \&c.$, until we get $AH = y$.

Moreover, take CD ∥ GK ∥ HL.

We have ((§ 23) X=AB:CD=CD:GK=GK: HL; and so

$$\frac{AB}{HL} = \left(\frac{AB}{CD}\right)^n$$

or $Y = X^n = X^{\frac{y}{x}}$.

If x, y are multiples of i, suppose $x = mi$, and $y = ni$; (by the preceding) $X = I^m$, $Y = I^n$, consequently

$$Y = X^{\frac{n}{m}} = X^{\frac{y}{x}}$$

The same is easily extended to the case of the incommensurability of x and y.

But if q=y−x, manifestly Q=Y:X.

It is also manifest that in Σ, for any x, we have X=1, but in S is X>1, and for any AB[11] and ABE there is such a CDF ‖ AB, that CDF =AB, whence AMBN≅AMEP, though the first be any multiple of the second; which indeed is singular, but evidently does not prove the absurdity of S.

§ 25. *In any rectilineal triangle, the circles with radii equal to its sides are as the sines of the opposite angles.*

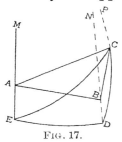

For take ∠ABC=rt.∠, and AM ⊥ BAC, and BN and CP ‖ AM; we shall have CAB ⊥ AMBN, and so (since CB ⊥ BA), CB ⊥ AMBN, consequently CPBN ⊥ AMBN.

Suppose the F of ray CP cuts the straights BN, AM respectively in D and E, and the bands CPBN, CPAM, BNAM along the L form lines CD, CE, DE. Then (§ 20) ∠CDE= the angle of NDC, NDE, and so =rt.∠; and by like reasoning ∠CED=∠CAB. But (by §21) in the L line △ CDE (supposing always here the radius =1),

EC:DC=1:sin DEC=1:sin CAB.

FIG. 17.

Also (by § 21)

EC:DC=⊙EC:⊙DC (in F)=⊙AC:⊙BC (§ 18);
and so is also

$$⊙AC:⊙BC=1:sin\ CAB;$$

whence the theorem is evident for any triangle.

§ 26. *In any spherical triangle, the sines of the sides are as the sines of the angles opposite.*

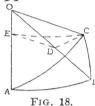

For take ∠ABC=rt.∠, and CED ⊥ to the radius OA of the sphere. We shall have CED ⊥ AOB, and (since also BOC ⊥ BOA), CD ⊥ OB. But in the triangles CEO, CDO (by § 25)

FIG. 18.

⊙EC:⊙OC:⊙DC=sin COE : 1 : sin COD=sin AC : 1 : sin BC; meanwhile also (§ 25) ⊙EC : ⊙DC=sin CDE : sin CED. Therefore, sin AC : sin BC=sin CDE : sin CED; but CDE= rt.∠=CBA, and CED=CAB. Consequently

$$sin\ AC : sin\ BC=1 : sin\ A.$$

Spherical trigonometry, flowing from this, is thus established independently of Axiom XI.

§ 27. If AC and BD are ⊥ AB, and CAB is carried along the straight AB; we shall have, designating by CD the path of the point C,

$$CD : AB=sin\ u : sin\ v.$$

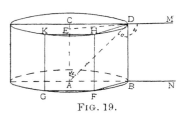

FIG. 19.

For take DE ⊥ CA; in the triangles ADE, ADB (by § 25)

$$\odot ED : \odot AD : \odot AB = \sin u : 1 : \sin v.$$

In revolving BACD about AC, B describes ⊙AB, and D describes ⊙ED; and designate here by s⊙CD the path of the said CD. Moreover, let there be any [12] polygon BFG... inscribed in ⊙AB.

Passing through all the sides BF, FG, &c., planes ⊥ to ⊙AB we form also a polygonal figure of the same number of sides in s⊙CD, and we may demonstrate, as in § 23, that CD : AB =DH : BF=HK : FG, &c., and so

DH+HK &c. : BF+FG &c. : =CD : AB.

If each of the sides BF, FG... approaches the limit zero, manifestly

BF+FG+...≐⊙AB   and
DH+HK+...≐⊙ED.

Therefore also ⊙ED : ⊙AB=CD : AB.   But we had ⊙ED : ⊙AB=sin u : sin v.   Consequently

CD : AB=sin u : sin v.

If AC goes away from BD to infinity, CD : AB, and so also sin u : sin v remains *constant;* but u≐rt. ∠ (§ 1), and if DM ‖ BN, v≐z; whence CD : AB=1 : sin z.

The path called CD will be denoted by CD ⫴ AB.

§ **28.** If BN ⫲ ≏ AM, and C in ray AM, and AC=x; we shall have (§ 23)

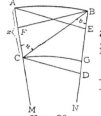

FIG. 20.

X=sin u : sin v.

For if CD and AE are ⊥ BN, and BF ⊥ AM; we shall have (as in § 27)

⊙BF : ⊙DC=sin u : sin v.

But evidently BF=AE: therefore

⊙EA : ⊙CD=sin u : sin v.

But in the F form surfaces of AM and CM (cutting AMBN in AB and CG) (by § 21)

⊙EA : ⊙DC=AB : CG=X.

Therefore also

X=sin u : sin v.

§ **29.** If ∠BAM=rt.∠, and sect AB=y, and BN ⫴ AM, we shall have in S

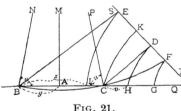

FIG. 21.

Y=cotan ½ u.

For, if sect AB= sect AC, and CP ⫴ AM (and so BN ⫲ ≏ CP), and ∠PCD= ∠QCD; there is given (§ 19) DS ⊥ ray CD, so that DS ⫴ CP, and so (§ 1) DT ⫴ CQ. Moreover, if BE ⊥ ray DS, then (§ 7) DS ⫴ BN, and so (§ 6)

BN ‖ ES, and (since DT ‖ CG) BQ ‖ ET; consequently (§ 1) ∠EBN=∠EBQ. Let BCF be
an L-line of BN, and FG, DH, CK, EL, L form
lines of FT, DT, CQ and ET; evidently (§ 22)
HG=DF=DK=HC; therefore,

$$CG=2CH=2v.$$

Likewise it is evident BG=2BL=2z.

But BC=BG−CG; wherefore $y=z-v$, and
so (§ 24) Y=Z : V.

Finally (§ 28)

$$Z=1 : \sin \tfrac{1}{2} u,$$
$$\text{and } V=1 : \sin (\text{rt.}\angle - \tfrac{1}{2} u),$$

consequently Y=cotan ½ u.

§ 30. However, it is easy to see (by § 25) [13]
that the solution of the problem of Plane

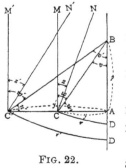

FIG. 22.

Trigonometry, in S, requires
the expression of the circle
in terms of the radius; but
this can by obtained by the
rectification of L.

Let AB, CM, C′M′ be ⊥
ray AC, and B anywhere in
ray AB; we shall have (§ 25)

$$\sin u : \sin v = \odot p : \odot y,$$
$$\text{and } \sin u' : \sin v' = \odot p' : \odot y';$$

and so $\dfrac{\sin u}{\sin v}.\odot y = \dfrac{\sin u'}{\sin v'}.\odot y'$.

But (by § 27) $\sin v : \sin v' = \cos u : \cos u'$;

consequently $\dfrac{\sin u}{\cos u} . \odot y = \dfrac{\sin u'}{\cos u'} . \odot y'$;

or $\odot y : \odot y' = \tan u' : \tan u = \tan w : \tan w'$.

Moreover, take CN and C'N' ‖ AB, and CD, C'D' L-form lines ⊥ straight AB; we shall have also (§21)

$$\odot y : \odot y' = r : r', \text{ and so}$$
$$r : r' = \tan w : \tan w'.$$

Now let p beginning from A increase to infinity; then $w \doteq z$, and $w' \doteq z'$, whence also $r : r' = \tan z : \tan z'$.

Designate by i the *constant*

$$r : \tan z \ (independent \text{ of } r);$$

whilst $y \doteq 0$,

$$\frac{r}{y} = \frac{i \tan z}{y} \doteq 1, \text{ and so}$$

$\dfrac{y}{\tan z} \doteq i.$ From § 29, $\tan z = \frac{1}{2} (Y - Y^{-1})$;

therefore $\dfrac{2y}{Y - Y^{-1}} \doteq i,$

or (§ 24) $\dfrac{2y . I^{\frac{y}{i}}}{I^{\frac{2y}{i}} - 1} \doteq i.$

But we know the limit of this expression (where $y \doteq 0$) is

$$\frac{i}{\text{nat. log I}}. \quad \text{Therefore}$$

$$\frac{i}{\text{nat. log I}}=i, \text{ and}$$
$$\text{I}=e=2.7182818\ldots,$$

which noted quantity shines forth here also.

If obviously henceforth i denote that sect of which the $\text{I}=e$, we shall have

$$r=i \tan z.$$

But (§ 21) $\odot y=2\pi r;$ therefore

$$\odot y=2\pi i \tan z=\pi i \ (\text{Y}-\text{Y}^{-1})=\pi i \left[e^{\frac{y}{i}}-e^{\frac{-y}{i}} \right]$$

$$=\frac{\pi y}{\text{nat. log Y}}(\text{Y}-\text{Y}^{-1}) \text{ (by § 24)}.$$

§ 31. For the trigonometric solution of all right-angled rectilineal *triangles* (whence the resolution of all *triangles* is easy , in S, three [14] equations suffice: indeed (a, b denoting the sides, c the hypothenuse, and a, β the angles opposite the sides) an equation expressing the relation

1st, between a, c, a;

2d, between a, a, β;

3d, between a, b, c;

of course from these equations emerge three others by elimination.

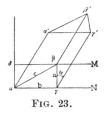

FIG. 23.

From §§ 25 and 30

$$1 : \sin a=(\text{C}-\text{C}^{-1}) : (\text{A}-\text{A}^{-1})=$$
$$= \left[e^{\frac{c}{i}}-e^{\frac{-c}{i}} \right] : \left[e^{\frac{a}{i}}-e^{\frac{-a}{i}} \right] \text{ (equation for } c, a \text{ and } a\text{)}.$$

II.   From § 27 follows (if βM ‖ γN)

cos a : sin β = 1 : sin u; but from § 29

1 : sin u = ½ (A+A^{-1});

therefore cos a sin β = ½ (A+A^{-1}) = ½ $\left(e^{\frac{a}{i}} + e^{\frac{-a}{i}} \right)$

(equation for a, β and a).

III.   If aa' ⊥ $\beta a\gamma$, and $\beta\beta'$ and $\gamma\gamma'$ ‖ aa' (§ 27),

and $\beta'a'\gamma'$ ⊥ aa'; manifestly (as in § 27)

$$\frac{\beta\beta'}{\gamma\gamma'} = \frac{1}{\sin u} = \tfrac{1}{2}(A+A^{-1});$$

$$\frac{\gamma\gamma'}{aa'} = \tfrac{1}{2}(B+B^{-1}\ ;$$

and $\dfrac{\beta\beta'}{aa'} = \tfrac{1}{2}(C+C^{-1})$; consequently

½ (C+C^{-1}) = ½ (A+A^{-1}). ½ (B+B^{-1}), or

$$\left(e^{\frac{c}{i}} + e^{\frac{-c}{i}} \right) = \tfrac{1}{2} \left(e^{\frac{a}{i}} + e^{\frac{-a}{i}} \right) \left(e^{\frac{b}{i}} + e^{\frac{-b}{i}} \right)$$

(equation for a, b and c).

If $\gamma a\delta$ = rt. ∠, and $\beta\,\delta$ ⊥ $a\delta$;

⊙c : ⊙a = 1 : sin a, and

⊙c : ⊙(d = $\beta\delta$) = 1 : cos a,

and so (denoting by ⊙x^2, for any x, the product ⊙x.⊙x) manifestly

$$\odot a^2 + \odot d^2 = \odot c^2.$$

But (by § 27 and II)

⊙d = ⊙b.½ (A+A^{-1}), consequently

$$\left(e^{\frac{c}{i}} - e^{\frac{-c}{i}} \right)^2 = \tfrac{1}{4} \left(e^{\frac{a}{i}} + e^{\frac{-a}{i}} \right)^2 \cdot \left(e^{\frac{b}{i}} - e^{\frac{-b}{i}} \right)^2 + \left(e^{\frac{a}{i}} - e^{\frac{-a}{i}} \right)^2$$

another equation for a, b and c (the second

member of which may be easily reduced to a
form *symmetric* or *invariable*).   [15]

Finally, from

$\dfrac{\cos a}{\sin \beta}=\tfrac{1}{2}(A+A^{-1})$, and $\dfrac{\cos \beta}{\sin a}=\tfrac{1}{2}(B+B^{-1})$, we get

(by III)

$$\cot a \cot \beta=\tfrac{1}{2}\left(e^{\frac{c}{i}}+e^{\frac{-c}{i}} \right)$$

(equation for a, β, and c.

§ **32.** It still remains to show briefly the
mode of resolving *problems* in S, which being
accomplished (through the more obvious examples), finally will be candidly said what this
theory shows.

I.   Take AB a line in a plane, and $y=f(x)$

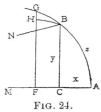

FIG. 24.

its equation in rectangular coordinates, call dz any increment
of z, and respectively dx, dy, du
the increments of x, of y, and of
the area u, corresponding to
this $dz;$ take BH ∥ CF, and ex-
press (from § 31) $\dfrac{\mathrm{BH}}{dx}$ by means of y, and seek

the *limit* of $\dfrac{dy}{dx}$ when dx tends towards the
limit zero (which is understood where a limit
of this sort is sought): then will become known
also the limit of $\dfrac{dy}{\mathrm{BH}}$, and so tan HBG; and

(since HBC manifestly is neither $>$ nor $<$, and so $=$rt. $\angle$), the *tangent* at B of BG will be determined by y.

II.   It can be demonstrated

$$\frac{dz^2}{dy^2 + \overline{BH}^2} \stackrel{\cdot}{=} 1.$$

Hence is found the *limit* of $\dfrac{dz}{dx}$, and thence, by integration, z (expressed in terms of x.

And of any line *given in the concrete,* the equation in S can be found; e. g., of L.   For if ray AM be the axis of L; then any ray CB from ray AM cuts L [since (by § 19) any straight from A except the straight AM will cut L]; but (if BN is axis)

X$=$1$:$sin CBN (§ 28),

and Y$=$cotan ½ CBN (§ 29), whence

Y$=$X$+\sqrt{X^2-1}$.

or $\qquad\qquad e^{\frac{y}{i}} = e^{\frac{x}{i}} + \sqrt{e^{\frac{2x}{i}} - 1},$ $\qquad$ [16]

the equation sought.

Hence we get

$$\frac{dy}{dx} \stackrel{\cdot}{=} X(X^2 - 1)^{-\frac{1}{2}};$$

and $\dfrac{BH}{dx} \stackrel{\cdot}{=} 1 : $ sin CBN$=$X; and so

$$\frac{dy}{BH} \stackrel{\cdot}{=} (X^2 - 1)^{-\frac{1}{2}};$$

$$1+\frac{dy^2}{BH^2}=X^2(X^2-1)^{-1},$$

$$\frac{dz^2}{BH^2}=X^2(X^2-1)^{-1},$$

and $\frac{dz}{BH}=X(X^2-1)^{-\frac{1}{2}}$, and

$\frac{dz}{dx}=X^2(X^2-1)^{-\frac{1}{2}}$, whence, by inte-

gration, we get (as in § 30)

$$z=i(X^2-1)^{\frac{1}{2}}=i \cot CBN.$$

III.   Manfestly

$$\frac{du}{dx}=\frac{HFCBH}{dx},$$

which (unless given in y) now first is to be expressed in terms of y; whence we get u by
integrating.

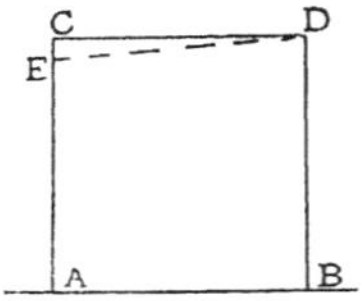

FIG. 25.

If AB=p, AC=q, CD=r, and
CABDC=s; we might show (as
in II) that
$\frac{ds}{dq}=r$, which $=\frac{1}{2}p\left(e^{\frac{q}{i}}-e^{\frac{-q}{i}}\right)$,

and, integrating, $s=\frac{1}{2}pi\left(e^{\frac{q}{i}}-e^{\frac{-q}{i}}\right)$

This can also be deduced apart from integration.

For example, the equation of the circle (from
§ 31, III), of the straight (from § 31, II), of a
conic (by what precedes), being expressed, the

areas bounded by these lines could also be expressed.

We know, that a surface t, ⫴ to a plane figure p (at the distance q), is to p in the ratio of the second powers of homologous lines, or as

$$\tfrac{1}{4}\left(e^{\frac{q}{i}}-e^{\frac{-q}{i}}\right)^2 : 1.$$

It is easy to see, moreover, that the calculation of volume, treated in the same manner, requires two integrations (since the differential itself here is determined only by integration); and before all must be investigated the [17] volume contained between p and t, and the aggregate of all the straights $\perp p$ and joining the boundaries of p and t.

We find for the volume of this solid (whether by integration or without it)

$$\tfrac{1}{8}pi\left(e^{\frac{2q}{i}}-e^{\frac{-2q}{i}}\right)+\tfrac{1}{2}pq.$$

The surfaces of bodies may also be determined in S, as well as the *curvatures,* the *involutes*, and *evolutes* of any lines, etc.

As to curvature; this in S either is the curvature of L, or is determined either by the radius of a circle, or by the *distance* to a straight from the curve ⫴ to this straight; since from what precedes, it may easily be shown, that in a plane there are no uniform lines other than L-lines, circles and curves ⫴ to a straight.

IV.   For the circle (as in III) $\dfrac{d \text{ area} \odot x}{dx} =$
$\odot x$, whence (by § 29), integrating,

$$\text{area } \odot x = \pi i^2 \left(e^{\frac{x}{i}} - 2 + e^{\frac{-x}{i}} \right).$$

V.   For the area CABDC$=u$ (inclosed by an

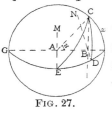

L form line AB$=r$, the ∥ to this, CD$=y$, and the sects AC$=$BD$=x$) $\dfrac{du}{dx} = y$; and (§ 24) $y = re^{\frac{-x}{i}}$, and so (integrating) $u = ri \left[1 - e^{\frac{-x}{i}} \right]$.

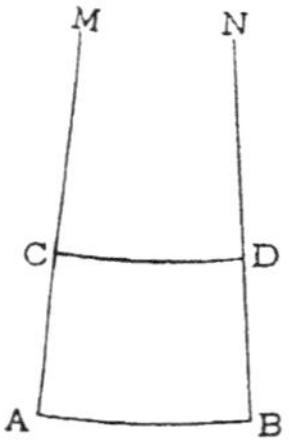

If x increases to infinity, then, in

FIG. 26.

S, $e^{\frac{-x}{i}} = 0$, and so $u = ri$. By the *size* of MABN, in future this limit is understood.

In like manner is found, if p is a figure on F, the space included by p and the aggregate of axes drawn from the boundaries of p is equal to ½pi.

VI.   If the angle at the center of a segment z of a sphere is $2u$, and a great circle is p, and x the arc FC (of the angle u); (§ 25)

FIG. 27.

$$1 : \sin u = p : \odot BC,$$
and hence $\odot BC = p \sin u$.   [18]

Meanwhile $x = \dfrac{pu}{2\pi}$, and $dx = \dfrac{pdu}{2\pi}$.

Moreover, $\frac{dz}{dx}=\odot BC$, and hence

$\frac{dz}{du}=\frac{p^2}{2\pi}\sin u$, whence (integrating)

$$z=\frac{\text{ver sin } u}{2\pi}p^2.$$

The F may be conceived on which P falls (passing through the middle F of the segment); through AF and AC the planes FEM, CEM are placed, perpendicular to F and cutting F along FEG and CE; and consider the L form CD (from C $\perp$ to FEG), and the L form CF; (§ 20) CEF=u, and (§ 21)

$\frac{FD}{p}=\frac{\text{ver sin } u}{2\pi}$, and so $z=FD.p$.

But (§ 21) $p=\pi.FGD$; therefore

$z=\pi.FD.FDG.$ But (§ 21)

M

$FD.FDG=FC.FC$; consequently

$z=\pi.FC.FC=\text{area } \odot FC$, in F.

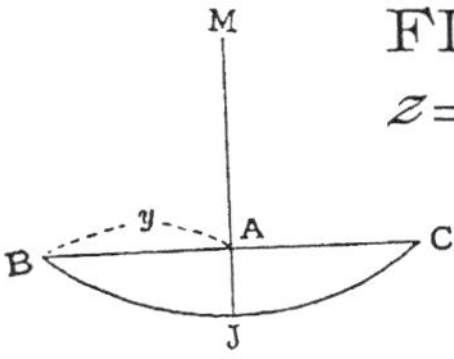

Now let BJ=CJ=r; (§ 30)

c $2r=i(Y-Y^{-1})$, and so (§ 21)

area $\odot 2r$ (in F) $=\pi i^2(Y-Y^{-1})^2$.

FIG. 28. Also (IV)

area $\odot 2y=\pi i^2(Y^2-2+Y^{-2})$;

therefore, area $\odot 2r$ (in F) =area $\odot 2y$, and so *the surface* z *of a segment of a sphere is equal to the surface of the circle described with the chord* FC *as a radius.*

Hence the whole surface of the sphere
$$=\text{area} \odot \text{FG} = \text{FDG}.p = \frac{p^2}{\pi},$$
and the surfaces of spheres are to each other as the second powers of their great circles.

VII. In like manner, in S, the volume of the sphere of radius x is found

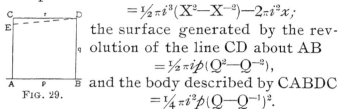

$$= \tfrac{1}{2}\pi i^3(X^2 - X^{-2}) - 2\pi i^2 x;$$
the surface generated by the revolution of the line CD about AB
$$= \tfrac{1}{2}\pi i p(Q^2 - Q^{-2}),$$
and the body described by CABDC
$$= \tfrac{1}{4}\pi i^2 p(Q - Q^{-1})^2.$$

FIG. 29.

But in what manner all things treated from (IV) even to here, also may be reached apart from integration, for the sake of brevity is suppressed.

It can be demonstrated that *the limit of every expression containing the letter i* (and so resting upon the hypothesis that i is given), [19] *when i increases to infinity, expresses the quantity simply for* $\mathfrak{L}$ (and so for the hypothesis of no i), if indeed the equations do not become identical.

But beware lest you understand to be supposed, that the system itself may be varied (for it is entirely determined in itself and by itself); but only *the hypothesis,* which may be

done successively, as long as we are not con-
ducted to an absurdity. *Supposing* therefore
that, in *such* an expression, the letter *i*, in
case S is reality, designates that unique quantity whose I=*e;* but if Σ is actual, the said
limit is supposed to be taken in place of the
expression: manifestly *all the expressions or-
iginating from the hypothesis of the reality
of S* (*in this sense*) *will be true absolutely,
although it be completely unknown whether
or not Σ is reality*

So e. g. from the expression obtained in § 30
easily (and as well by aid of differentiation as
apart from it) emerges the known value in Σ,
$$\odot x = 2\pi x;$$
from I (§ 31) suitably treated, follows
$$1 : \sin a = c : a;$$
but from II
$$\frac{\cos a}{\sin \beta} = 1, \text{ and so}$$
$$a + \beta = \text{rt.} \angle;$$
the first equation in III becomes identical, and
so is true in Σ, although it there determines
nothing; but from the second follows
$$c^2 = a^2 + b^2.$$
*These are the known fundamental equa-
tions of plane trigonometry in Σ.*

Moreover, we find (from § 32) in Σ, the area and the volume in III each $=pq;$ from IV

$$\text{area} \odot x = \pi x^2;$$

(from VII) the globe of radius x

$$= \tfrac{4}{3}\pi x^3, \text{ etc.}$$

The theorems enunciated at the end of VI are manifestly *true unconditionally*.

§ **33.** It still remains to set forth (as promised in § 32) what this theory means.

I.   Whether Σ or some one S is reality, remains undecided.

II.   All things deduced from the hypothesis of the falsity of Axiom *XI* (always to be understood in the sense of § 32) are *absolutely true,* and so in this sense, *depend upon no hypothesis.*

There is therefore *a plane trigonometry a priori, in which the system alone really re-* [20] *mains unknown;* and so where remain unknown solely the *absolute* magnitudes in the expressions, but where a *single* known case would manifestly fix the whole system.   But spherical trigonometry is established absolutely in § 26.

(And we have, on F, a geometry wholly analogous to the plane geometry of Σ.)

III.   If it were agreed that Σ exists, nothing more would be unknown in this respect; but

if it were *established* that Σ *does not exist*,
then (§ 31), (e. g.) from the sides *x, y,* and the
rectilineal angle they include being given in a
special case, manifestly it would be impossible
in itself and by itself to solve absolutely the
triangle, that is, to determine *a priori* the
other angles and *the ratio of the third side* to
the two given; unless X, Y were determined,
for which it would be necessary to have in
concrete form a certain sect *a* whose A was
known; and then *i* would be *the natural unit
for length* (just as *e* is the base of *natural*
logarithms).

If the existence of this *i* is determined, it
will be evident how it could be constructed,
at least very exactly, for practical use.

IV. In the sense explained (I and II), it is
evident that all things in space can be solved
by the modern analytic method (within just
limits strongly to be praised).

V. Finally, to friendly readers will not be
unacceptable; that for that case wherein not Σ
but S is reality, a rectilineal figure is constructed equivalent to a circle.

§ 34. Through D we may draw DM ‖ AN in
the following manner. From D drop DB⊥AN;
from any point A of the straight AB erect AC
⊥ AN (in DBA), and let fall DC⊥AC. We

will have (§ 27) $\odot CD : \odot AB = 1 : \sin z$, pro-vided that DM ‖ BN. But sin z is not >1; and so AB is not $>$ DC. Therefore a quadrant described from the center A in BAC, with a radius $=$ DC, will have a point B or O in common with ray BD. In the first case, manifestly $z = $ rt. $\angle$; but in the second case (§ 25)

$$(\odot AO = \odot CD) : \odot AB = 1 : \sin AOB,$$

and so $z = AOB$.

If therefore we take $z = AOB$, then DM will be ‖ BN.

§ **35.** If S were reality; we may, as follows, draw a straight $\perp$ to one arm of an acute angle, [21] which is ‖ to the other.

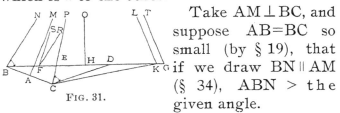

FIG. 31.

Take AM $\perp$ BC, and suppose AB $=$ BC so small (by § 19), that if we draw BN ‖ AM (§ 34), ABN $>$ the given angle.

Moreover draw CP ‖ AM (§ 34); and take NBG and PCD each equal to the given angle; rays BG and CD will cut; for if ray BG (falling *by construction* within NBC) cuts ray CP in E; we shall have (since BN $\backsimeq$ CP), $\angle$EBC$<$ $\angle$ECB, and so EC$<$EB. Take EF$=$EC, EFR

=ECD, and FS ‖ EP; then FS will fall within BFR. For since BN ‖ CP, and so BN ‖ EP, and BN ‖ FS; we shall have (§ 14)

∠FBN+∠BFS< (st.∠=FBN+BFR);

therefore, BFS<BFR. Consequently, ray FR cuts ray EP, and so ray CD also cuts ray EG in some point D. Take now DG=DC and DGT=DCP=GBN; we shall have (since CD ≏ GD) BN≏GT≏CP. Let K (§ 19) be the point of the L-form line of BN falling in the ray BG, and KL the axis; we shall have BN≏KL, and so BKL=BGT=DCP; but also KL≏CP: therefore manifestly K fall on G, and GT ‖ BN. But if HO bisects ⊥ BG, we shall have constructed HO ‖ BN.

§ 36. Having given the ray CP and the plane MAB, take CB ⊥ the plane MAB, BN (in plane BCP) ⊥ BC, and CQ ‖ BN (§ 34); the intersection of ray CP (if this ray falls within BCQ) with ray BN (in the plane CBN), and so with the plane MAB is found. And if we are given the two planes PCQ, MAB, and we have CB ⊥ to plane MAB, CR ⊥ plane PCQ; and (in plane BCR) BN⊥BC, CS⊥CR, BN will fall in plane MAB, and CS in plane PCQ; and the

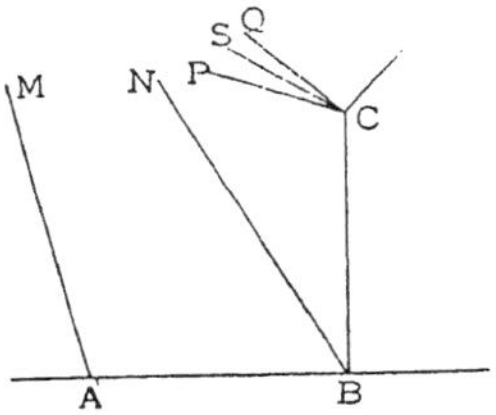

Fig. 32.

intersection of the straight BN with the straight CS (if there is one) having been found, the perpendicular drawn through this intersection, in PCQ, to the straight CS will manifestly be the intersection of plane MAB and plane PCQ.

§ 37. On the straight AM ‖ BN, is found such

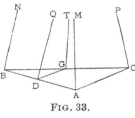

FIG. 33.

an A, that AM ≏ BN. If (by [22] § 34) we construct outside of the plane NBM, GT ‖ BN, and make BG⊥GT, GC=GB, and CP‖GT; and so place the hemiplane TGD that it makes with hemi-plane TGB an angle equal to that which hemi-plane PCA makes with hemi-plane PCB; and is sought (by § 36) the intersection straight DQ of hemi-plane TGD with hemiplane NBD; and BA is made ⊥ DQ.

We shall have indeed, on account of the similitude of the triangles of L lines produced on the F of BN (§ 21), manifestly DB=DA, and AM ≏ BN.

Hence easily appears (L-lines being given by their extremities alone) we may also find a fourth proportional, or a mean proportional, and execute in this way in F, apart from Axiom XI, all the geometric constructions made

on the plane in Σ. Thus e. g. a perigon can
be geometrically divided into any special number of equal parts, if it is permitted to make
this special partition in Σ.

§ 38. If we construct (by § 37) for example,

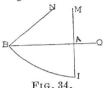

NBQ=⅓ rt.∠, and make (by
§ 35), in S, AM⊥ray BQ and ‖
BN, and determine (by § 37)
IM⇌BN; we shall have, if IA

Fig. 34. $=x$, (§ 28), X=1 : sin ⅓ rt. ∠=2,
and x will be constructed *geometrically.*

And NBQ may be so computed, that IA differs from i less than by anything given, which
happens for sin NBQ=$^1/e$.

§ 39. If (in a plane) PQ and ST are ‖ to the
straight MN (§ 27), and AB, CD are equal
perpendiculars to MN; manifestly △DEC≅

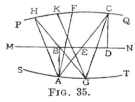

△BEA; and so the angles
(perhaps mixtilinear) ECP,
EAT will fit, and EC=EA.
If, moreover, CF=AG, then
△ACF≅△CAG, and each
is half of the *quadrilateral*

FAGC.

If FAGC, HAGK are two quadrilaterals of
this sort on AG, between PQ and ST; their
equivalence (as in Euclid) is evident, as also

the equivalence of the triangles AGC, AGH, standing on the same AG, and having their vertices on the line PQ. Moreover, ACF= CAG, GCQ=CGA, and ACF+ACG+GCQ= st. ∠ (§ 32); and so also CAG+ACG+CGA= [23] st. ∠; therefore, in any triangle ACG of this sort, the sum of the three angles =st. ∠. But whether the straight AG may have fallen upon AG (which ‖ MN), or not; *the equivalence* of the rectilineal triangles AGC, AGH, as well of themselves, *as of the sums of their angles,* is evident.

§ 40. *Equivalent triangles* ABC, ABD, (henceforth rectilineal), *having one side equal, have the sums of their angles equal.*

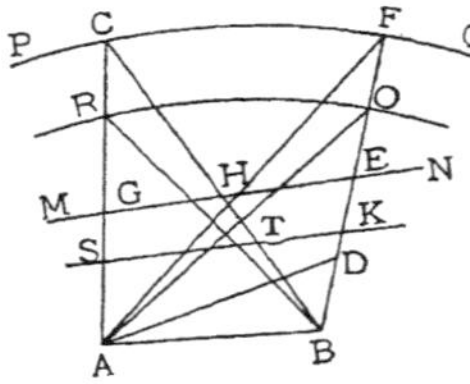

FIG. 36.

For let MN bisect AC and BC, and take (through C) PQ ‖ MN; the point D will fall on line PQ.

For, if ray BD cuts the straight MN in the point E, and so (§ 39) the line PQ at the distance EF=EB; we shall have △ABC=△ABF, and so also △ABD=△ABF, whence D falls at F.

But if ray BD has not cut the straight MN, let C be the point, where the perpendicular bisecting the straight AB cuts the line PQ, and

let GS=HT, so, that the line ST meets the
ray BD prolonged in a certain K (which it is
evident can be made in a way like as in § 4);
moreover take SR=SA, RO ⫴ ST, and O the
intersection of ray BK with RO; then △ABR
=△ABO (§ 39), and so △ABC>△ABD (contra hyp.).

§ **41.** *Equivalent triangles ABC, DEF
have the sums of their triangles equal.*

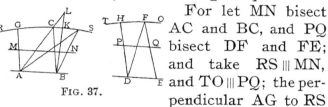

For let MN bisect
AC and BC, and PQ
bisect DF and FE;
and take RS ⫴ MN,
and TO ⫴ PQ; the per-
pendicular AG to RS

FIG. 37.

will equal the perpendicular DH to TO, or one
for example DH will be the greater.

In each case, the ⊙DF, from center A, has
with line-ray GS some point K in common,
and (§ 39) △ABK=△ABC=△DEF. But the
△AKB (by § 40) has the same angle-sum as
△DFE, and (by § 39) as △ABC. Therefore
also the triangles ABC, DEF have each the
same angle-sum.

In S the inverse of this theorem is true.

For take ABC, DEF two triangles having
equal angle-sums, and △BAL=△DEF; these
will have (by what precedes) equal angle-sums,

and so also will $\triangle ABC$ and $\triangle ABL$, and hence manifestly

$$BCL+BLC+CBL=\text{st.}\angle.$$

However (by § 31), the angle-sum of any tri- [24] angle, in S, is $<\text{st.}\angle$.

Therefore L falls on C.

§ **42.** Let u be the supplement of the angle-sum of the $\triangle ABC$, but v of $\triangle DEF$; then is $\triangle ABC : \triangle DEF = u : v$.

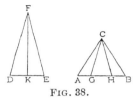

For if p be the area of each of the triangles ACG, GCH, HCB, DFK, KFE; and $\triangle ABC = m.p$, and $\triangle DEF = n.p;$ and s the angle-sum of any triangle equivalent to $p;$

FIG. 38.

manifestly

st. $\angle - u = m.s - (m-1)$st. $\angle = $st. $\angle - m($st. $\angle - s);$
and $u = m($st. $\angle - s);$ and in like manner $v = n($st. $\angle - s).$

Therefore $\triangle ABC : \triangle DEF = m : n = u : v.$

It is evidently also easily extended to the case of the incommensurability of the triangles ABC, DEF.

In the same way is demonstrated that triangles on a sphere are as the *excesses* of the sums of their angles above a st. $<$.

If two angles of the spherical $\triangle$ are right, the third z will be the said *excess*. But (a great circle being called p) this $\triangle$ is manifestly

$$=\frac{z}{2\pi}\,\frac{p^2}{2\pi}\ (\S\ 32,\ \mathrm{VI});$$

consequently, any triangle of whose angles the excess is z, is

$$=\frac{zp^2}{4\pi^2}.$$

§ 43. Now, in S, the area of a rectilineal $\triangle$ is expressed by means of the sum of its angles.

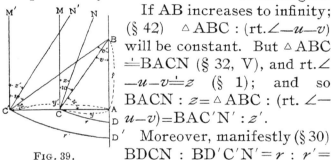

Fig. 39.

If AB increases to infinity; (§ 42) $\triangle\,\mathrm{ABC} : (\mathrm{rt.}\angle-u-v)$ will be constant. But $\triangle\,\mathrm{ABC}$ $\doteqdot\mathrm{BACN}$ (§ 32, V), and rt.$\angle$ $-u-v\doteqdot z$ (§ 1);   and   so $\mathrm{BACN} : z = \triangle\,\mathrm{ABC} : (\mathrm{rt.}\ \angle-u-v)=\mathrm{BAC'N'} : z'$.

Moreover, manifestly (§ 30) $\mathrm{BDCN} : \mathrm{BD'C'N'}=r : r'=$ $\tan z : \tan z'$.

But for $y'\doteqdot 0$, we have $$\frac{\mathrm{BD'C'N'}}{\mathrm{BAC'N'}}\doteqdot 1,\text{ and also }\frac{\tan z'}{z'}\doteqdot 1;$$

consequently,

$$\mathrm{BDCN} : \mathrm{BACN}=\tan z : z.$$

But (§ 32)

$$\mathrm{BDCN}=r.i=i^2\tan z;$$

therefore, $\mathrm{BACN}=z.i^2$.

Designating henceforth, for brevity, any triangle the supplement of whose angle-sum is z by $\triangle$, we will therefore have $\triangle = z.i^2$.

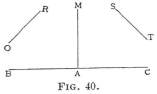

FIG. 40.

Hence it readily flows that, if OR‖AM and RO‖AB, the *area* comprehended between the straights OR, ST, BC [25] (which is manifestly the absolute limit of the area of rectilineal triangles increasing without bound, or of $\triangle$ for $z \doteq$ st. $\angle$), is $= \pi i^2 =$ area $\odot i$, in F.

This limit being denoted by $\square$, moreover (by § 30) $\pi r^2 = \tan^2 z. \square =$ area $\odot r$ in F (§ 21) = area $\odot s$ (by §32, VI) if the chord CD is called s.

If now, bisecting at right angles the given radius s of the circle in a plane (or the L form radius of the circle in F), we construct (by § 34) DB‖⇌CN; by dropping CA ⊥ DB, and erecting CM ⊥ CA, we shall

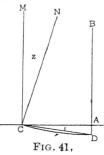

FIG. 41,

get z; whence (by § 37), assuming at pleasure an L form radius for unity, $\tan^2 z$ *can be determined geometrically by means of two uniform lines of the same curvature* (which, their extremities alone being given and their axes constructed, manifestly may be compared like straights, and in this respect considered equivalent to straights).

Moreover, a quadrilateral, ex. gr. regular = □ is constructed as follows:

FIG. 42.

Take ABC=rt.∠, BAC=½ rt. ∠, ACB=¼ rt. ∠, and BC=x.

By mere square roots, X (from § 31, II) can be expressed and (by § 37) constructed; and having X (by § 38 or also §§ 29 and 35), x itself can be determined. And octuple △ABC is manifestly = □, and by this *a plane circle of radius s is geometrically squared by means of a rectilinear figure and uniform lines of the same species* (equivalent to straights as to comparison *inter se*); *but an F form circle is planified in the same manner: and we have either the Axiom XI of Euclid true or the geometric quadrature of the circle,* although thus far it has remained undecided, which of these two has place in reality.

Whenever tan²z is either a whole number, or a rational fraction, whose denominator (reduced to the simplest form) is either a prime number of the form 2^m+1 (of which is also $2=2^0+1$), or a product of however many prime numbers of this form, of which each (with the

exception of 2, which alone may occur any
number of times) occurs *only once* as factor,
we can, by the theory of polygons of the illus-
trious Gauss (remarkable invention of our,
nay of every age) (and only for such values [26]
of z), construct a rectilineal figure $= \tan^2 z \, \square =$
area $\odot s$. For the division of $\square$ (the theorem
of § 42 extending easily to any polygons) manifestly requires the partition of a st. $\angle$, which
(as can be shown) can be achieved geometrically only under the said condition.

But in all such cases, what precedes con-
ducts easily to the desired end. And any rectilineal figure can be converted geometrically
into a regular polygon of n sides, if n falls
under the Gaussian form.

It remains, finally (that the thing may be
completed in every respect), to demonstrate
the impossibility (apart from any supposition),
of deciding *a priori*, whether Σ, or some S
(and which one) exists. This, however, is re-
served for a more suitable occasion.

APPENDIX I.

REMARKS ON THE PRECEDING TREATISE,
BY BOLYAI FARKAS.

[From Vol. II of Tentamen, pp. 380–383.]

Finally it may be permitted to add something appertaining to the author of the *Appendix* in the first volume, who, however, may pardon me if something I have not touched with his acuteness.

The thing consists briefly in this: *the formulas of spherical trigonometry* (demonstrated in the said *Appendix* independently of Euclid's Axiom XI) *coincide with the formulas of plane trigonometry, if* (in a way provisionally speaking) *the sides of a spherical triangle are accepted as reals, but of a rectilineal triangle as imaginaries;* so that, as to trigonometric formulas, the plane may be considered as an imaginary sphere, if for real, that is accepted in which sin rt. $\angle = 1$.

Doubtless, of the Euclidean axiom has been said in volume first enough and to spare: for

the case if it were not true, is demonstrated
(Tom. I. App., p. 13), that there is given a certain i, for which the I there mentioned is $=e$
(the base of natural logarithms), and for this
case are established also (*ibidem*, p. 14) the
formulas of plane trigonometry, and indeed so,
that (by the side of p. 19, ibidem) the formulas
are still valid for the case of the verity of the
said axiom; indeed if the limits of the values
are taken, supposing that $i \doteq \infty$; truly the
Euclidean system is as if the limit of the anti-
Euclidean (for $i \doteq \infty$).

Assume for the case of i existing, the unit
$= i$, and extend the concepts sine and cosine
also to imaginary arcs, so that, p designating
an arc whether real or imaginary,

$$\frac{e^{p\sqrt{-1}} + e^{-p\sqrt{-1}}}{2} \text{ is called the}$$

cosine of p, and

$$\frac{e^{p\sqrt{-1}} - e^{-p\sqrt{-1}}}{2\sqrt{-1}} \text{ is called}$$

the *sine* of p (as Tom. I., p. 177).

Hence for q real

$$\frac{e^q - e^{-q}}{2\sqrt{-1}} = \frac{e^{-q\sqrt{-1}\cdot\sqrt{-1}} - e^{q\sqrt{-1}\cdot\sqrt{-1}}}{2\sqrt{-1}} = \sin(-q\sqrt{-1})$$
$$= -\sin(q\sqrt{-1}).$$

So $\dfrac{e^{q}+e^{-q}}{2}=\dfrac{e^{-q\sqrt{-1}.\sqrt{-1}}+e^{q\sqrt{-1}.\sqrt{-1}}}{2}=\cos(-q\sqrt{-1})$
$$=\cos(q\sqrt{-1});$$

if of course also in the imaginary circle, the
sine of a negative arc is the same as the sine
of a positive arc otherwise equal to the first,
except that it is negative, and the cosine of a
positive arc and of a negative (if otherwise
they be equal) the same.

In the said *Appendix*, § 25, is demonstrated
absolutely, that is, independently of the said
axiom; that, in any rectilineal triangle *the
sines of the circles are as the circles of radii
equal to the sides opposite.*

Moreover is demonstrated for the case of *i*
existing, that the circle of radius *y* is

$$=\pi i\ \left[e^{\frac{y}{i}}-e^{\frac{-y}{i}}\right], \text{ which, for } i=1, \text{ becomes}$$
$$\pi(e^{y}-e^{-y}).$$

Therefore (§ 31 *ibidem*), for a right-angled
rectilineal triangle of which the sides are *a*
and *b*, the hypothenuse *c*, and the angles opposite to the sides *a, b, c* are *α, β*, rt. $\angle$, (for $i=1$),
in I,

$$1:\sin\ a=\pi(e^{c}-e^{-c}):\pi(e^{a}-e^{-a});$$

and so

$$1:\sin a =\frac{e^{c}-e^{-c}}{2\sqrt{-1}}:\frac{e^{a}-e^{-a}}{2\sqrt{-1}}. \qquad \text{Whence } 1:\sin a$$

$= -\sin\,(c\sqrt{-1}) : -\sin\,(a\sqrt{-1}).$ And hence

$1 : \sin\,\alpha = \sin\,(c\sqrt{-1}) : \sin\,(a\sqrt{-1}).$

In II becomes

$\cos\,\alpha : \sin\,\beta = \cos\,(a\sqrt{-1}) : 1;$

in III becomes

$\cos\,(c\sqrt{-1}) = \cos\,(a\sqrt{-1}).\cos\,(b\sqrt{-1}).$

These, as all the formulas of plane trigonometry deducible from them, coincide completely with the formulas of spherical trigonometry; except that if, ex. gr., also the sides and the angles opposite them of a right-angled spherical triangle and the hypothenuse bear the same names, the sides of the rectilineal triangle are to be divided by $\sqrt{-1}$ to obtain the formulas for the spherical triangle.

Obviously we get (clearly as Tom., II., p. 252),

from I, $1 : \sin\,\alpha = \sin\,c : \sin\,a;$

from II, $1 : \cos\,a = \sin\beta : \cos\,\alpha;$

from III, $\cos\,c = \cos\,a\,\cos\,b.$

Though it be allowable to pass over other things; yet I have learned that the reader may be offended and impeded by the deduction omitted, (Tom. I., App., p. 19) ⌊in § 32 at end⌋: it will not be irrelevant to show how, ex. gr., from

$$e^{\frac{c}{i}} + e^{\frac{-c}{i}} = \tfrac{1}{2}\left(e^{\frac{a}{i}} + e^{\frac{-a}{i}}\right)\left(e^{\frac{b}{i}} + e^{\frac{-b}{i}}\right)$$

follows

$$c^2=a^2+b^2.$$

(the theorem of Pythagoras for the Euclidean system); probably thus also the author deduced it, and the others also follow in the same manner.

Obviously we have, the powers of e being expressed by series (like Tom. I., p. 168),

$$e^{\frac{k}{i}}=1+\frac{k}{i}+\frac{k^2}{2i^2}+\frac{k^3}{2.3.i^3}+\frac{k^4}{2.3.4.i^4} \cdots \cdots,$$

$$e^{\frac{k}{i}}=1-\frac{k}{i}+\frac{k^2}{2i^2}-\frac{k^3}{2.3.i^3}+\frac{k^4}{2.3.4.i^4} \cdots \cdots, \text{ and so}$$

$$e^{\frac{k}{i}}+e^{\frac{-k}{i}}=2+\frac{k^2}{i^2}+\frac{k^4}{3.4.i^4}+\frac{k^6}{3.4.5.6.i^6} \cdots \cdots,$$

$$=2+\frac{k^2+u}{i^2}, \text{ (designating by}$$

$\frac{u}{i^2}$ the sum of all the terms after $\frac{k^2}{i^2}$) ; and we have $u\doteq0$, while $i\doteq\infty$. For all the terms which follow $\frac{k^2}{i^2}$ are divided by i^2; the first term will be $\frac{k^4}{3.4i^2}$; and any ratio $<\frac{k^2}{i^2}$; and though the ratio everywhere should remain this, the sum would be (Tom. I., p. 131),

$$\frac{k^4}{3.4.i^2} : \left(1-\frac{k^2}{i^2}\right) =\frac{k^4}{3.4.(i^2-k^2)},$$

which manifestly $\doteq0$, while $i\doteq\infty$.

And from

$$e^{\frac{c}{i}} + e^{\frac{-c}{i}} = \tfrac{1}{2}\left(e^{\frac{(a+b)}{i}} + e^{\frac{-(a+b)}{i}} + e^{\frac{a-b}{i}} + e^{\frac{-(a-b)}{i}} \right)$$

follows (for w, v, λ taken like u)

$$2 + \frac{c^2 + w}{i^2} = \tfrac{1}{2}\left(2 + \frac{(a+b)^2 + v}{i^2} + 2 + \frac{(a+b)^2 + \lambda}{i^2} \right).$$

And hence

$$c^2 = \frac{a^2 + 2ab + b^2 + a^2 - 2ab + b^2 + v + \lambda - w}{2},$$

which $\overset{\text{.}}{=} a^2 + b^2$.

Library of Congress Cataloging-in-Publication Data

Gray, Jeremy, 1947–
   János Bolyai, non-Euclidean geometry, and the nature of space /
   Jeremy J. Gray.
       p. cm.–(Burndy Library publications ; new ser., no. 1)
   Includes bibliographical references.
   ISBN 0-262-57174-9 (pbk.)
   1. Geometry, Non-Euclidean. 2. Space and time. 3. Bolyai, János,
   1802–1860. I. Bolyai, János, 1802–1860. Appendix scientiam spatii
   absolute veram exhibens. English. II. Title. III. Series.

QA685.G73 2004
516.9–dc22
                                              2003071062

THE BURNDY LIBRARY was founded in 1936 by Bern Dibner, an electrical engineer and businessman who developed a great passion for the history of science and technology. The Library, now located on the campus of the Massachusetts Institute of Technology in Cambridge, Massachusetts, is one of the world's leading resources for the history of science and technology, housing a collection of over fifty thousand books and manuscripts that span history from antiquity to the present.

From 1941 until Bern Dibner's death in 1989, the Library issued occasional publications meant to highlight special treasures in the Library's collection and important developments in the history of science and technology. This volume is the first in a new series of such publications, intended, in the spirit of the original series, to be both useful to scholars and accessible to the nonspecialist.

PRINTED AT MERIDIAN PRINTING COMPANY,

EAST GREENWICH, RHODE ISLAND

ON MOHAWK SUPERFINE PAPER

TYPE COMPOSED IN ITC BODONI

DESIGNED BY SUSAN MARSH

1,500 COPIES